Crenel Physics

Part 1: Introducing the lean approach

by:

Hans van Kessel

Contents:

1. Introduction: 'Crenel Physics'.

In the Medieval centuries knights built castles from which they could conquer the world. Solid walls surrounding these castles ensured protection against intruders. The crenellation on top of these walls improved the effectiveness of the defense: soldiers could hide behind the elevated wall sections, and could shoot through the crenels -the lower sections- with minimal exposure.

Figure 1.1: Crenels on top of a castle wall.

The presence of crenellation characterizes a building as a castle.

In physics the shape of crenellation represents the *most basic form of change*, the so called *binary function*. On top of the castle wall, the upper section is followed by the

3

lower section, which is in turn followed by an upper section, and so on.

Because there are only two statuses, the walking along such a wall is like walking along a *binary* function. The two statuses of such a binary function can have more general names, like: 'True' and 'False', 'Yes' and 'No', '1' and '0', 'on' and 'off', etcetera. In concept, the names that are given to both statuses are not relevant, as long as there are just two options.

Based on a binary function, the answer to a question about the *status* of an object may neither always (and anywhere) be a 'True', nor always (and anywhere) be a 'False'. The binary function thus represents the *leanest possible* form of 'change'. And when graphically represented, it has the looks of crenellation on top of a castle wall. The crenellation's extremely basic shape thus is a metaphor for leanness, and inspired to the name 'Crenel Physics':

> *'Crenel Physics' is one effort to describe physics in the leanest possible way.*

Through the metaphor, the name 'Crenel Physics' can indeed be associated with the objective: ultimate leanness.

The objective requires the redefinition of current and commonly used 'metrics' for describing processes and phenomena. The used metrics should be such that only physical essences are expressed in a completely objective and unambiguous manner. Prior to that, it will be made clear that the historically grown and widely used metric

'SI' system is far from that. The effort to achieve this redefinition is referred to as 'normalization'.

Such efforts have already been ongoing for quite some time, primarily driven by theoretical physicists. Nevertheless: based on today's state of the art there is no unambiguous and generally agreed upon 'normalization' in physics. This is one indication that physicists are not done yet. Furthermore, there still is no complete insight into all fundamentals. This –of course- is challenging news. There still remains work to do…

After such a 'normalization' (as good as it can get), the current physical rules/relations will need translation, based on this new set of units. Finally and hopefully, new –more fundamental- interpretations may come forth from the effort. Fundamental insights typically lead to new results, because these insights are as close as possible to *the* 'objective physical reality' that lies at the basis of modern technology.

Many physicists argue, that communication with potential other intelligent life forms requires 'objective means' (or: 'normalized units'). There even is a possibility that even the *finding* of some other life forms requires 'universal signals'. These limitations would disqualify e.g. the use of sounds or colors as communication vehicle (as used in the movie 'Close encounters of the third kind' when finally contact is made…) because sounds and colors are based on human perceptions (of the non-deaf and non-color blind persons). From an objective viewpoint colors do not exist… Colors are a human interpretation of photon energy levels (see chapter 13).

It also is thinkable that the behavior of known species on Earth –such as horses- can be better understood if humans would take into account that human perceptions like 'sound' or 'color' are not always equally shared.

As mentioned, the search for a 'universally objective' set of units of measurement is an ongoing effort. In the 1880-ties the Irish physicist George Johnstone Stoney started to work his way back from the daily used metrics towards the fundamentals underneath. What he was looking for –and partially found- became later known as 'natural units'. The particular ones he proposed were –in his honor- referred to as 'Stoney units'. Thereafter, in the beginning of the 20th century, amongst others the German physicist Max Planck enhanced the process of normalization, based on the further developed level of fundamental knowledge. Thus, a set of 'Planck units' were defined. However, until there is a generally accepted 'theory of everything', one can assume that the goal has not been reached.

The searched for leanness (or: restriction to fundamental units only) is not the equivalent of *poverty*, nor is it limiting the complexity or beauty in outcomes.

For example: a series of the aforementioned extremely lean binary functions can be accumulated. These functions can be used as the elementary building blocks for more impressive changes. Albeit that in doing so:

- A (very) *large amount* of these building blocks may be required, and -due to the accumulation of a *discrete number* of these blocks-

6

- There is a build in limitation with regards to the *level of detail* (or smoothness) that possibly can be achieved.

At close look the -more complex- accumulated results would indeed and inherently have sharp angles. For the smoothening of these angles there is *no* such thing as e.g. '*half* an elementary building block'. This would be a 'contradictio interminis': *elementary* blocks cannot be cut into smaller pieces. Or: binary functions cannot be further simplified.

Despite this, modern computers nowadays demonstrate the capabilities that are offered by managing large quantities of these ultra lean binary building blocks (and processes): *any* result represented or performed through computers is produced by the managing of -large- amounts of binary changes. This is because computers are –nowadays- strictly binary machines. Computers can only *compare* or *add* bits (a binary numbers is a string of bits), and move the results thereof around in memory. Even a simple process like *multiplication* is not executed as such. Multiplication is performed through an algorithm: a –managed- series of binary bit by bit additions, and shifts of binary numbers in various memory locations. That's why computers are rightfully nicknamed 'fast idiots'.

Prior to computers, the world was much more envisioned as being analogue, as being non-discrete. The 'quantum' was hypothetical and initially it felt unnatural and insignificant to many scientists.

But even today, the here targeted *leanest possible* way to describe physics will require some major gymnastics of the human brain. This is because the human brain is adapted to a *perceived* (and analogue) world, and not to the fundamental world that is searched for here.

To illustrate the challenge at hand: the aforementioned binary change can only be envisioned through the human brain if it takes place *somewhere*, and *sometime*. This *somewhere* and *sometime* however are both human perceptions. They are *not* objective: the 'theory of relativity' says that the 'somewhere' and 'sometime' are *relative*. The human perceptions thus create a human reality that turned out to be *not* universally sharable without conversion. This is one indication that underneath this human reality a layer of more fundamental properties must exist. It is this more fundamental layer that is of interest here. In Crenel Physics, *the* most fundamental layer –at the basis of any physical property- is the target.

As discussed, this requires a search for the so called 'natural units' of measurement. From a historically grown human perspective these units will however not feel 'natural' at all. Rather than 'natural', the emotionally colder word 'fundamental' would be a much better terminology to use here. Unfortunately the word 'fundamental' has already been used by physicists too often and too soon in a non-fundamental context... and may therefore cause ambiguity. Many books about 'fundamental physics' in fact use the metric SI system for expressing units of measurement. These SI units are historically grown over centuries and generally are based on human arbitrariness. These books may indeed describe fundamental physics, but not by using fundamental –objective- terminology. These books would be of little value to Martians (other than serving the objective of studying and understanding Earthly humans).

To avoid potential confusion or misinterpretations, an intentionally exclusive terminology will be introduced and used in Crenel Physics. It will be based on the outcome of the 'redone from scratch' search for 'natural units'. This exclusive terminology itself will - from here onwards - be generally referred to as *Crenel Physics*. Fortunately, as a consequence to the intended lean approach, the number of new terms that thus will be introduced is limited to the extreme (in this book there will only be two).

This, as opposed to today's more common science in which indeed many properties, standards or units of measurement have been introduced historically, and are – at least partially- used routinely in daily life. These numerous units of measurement (or properties) contain their relativity, overlap and redundancy because they were historically introduced *before* the current insight into underlying fundamental physical properties matured. Some examples thereof will be given.

When referring to today's more common descriptions of science, the *metric* system (based on so called 'SI-units') will be used as the reference. As opposed to 'Crenel Physics', this more common way of describing today's science will generally be referred to as 'Metric Physics' (even though not all units may be part of the 'metric' SI system):

The term *Metric Physics*:
will be used to generally refer to the more common ways of describing physics, e.g. using SI-units of measurement.

If *Metric Physics* indeed contains too many physical properties and standards, why has it not been cleaned out by now? The answer is: this would be too much of a challenge for one generation of humans. Many of these properties and standards historically became part of daily life and thinking. And -once firmly settled- the human brain has great difficulty in setting these overboard. For example: we humans keep using the terms 'yardstick' and 'milestone', even where metric units are the standard. 'Yards' and 'miles' are no part of that.

Under the circumstances, *any* initiative to gradually push science into the direction of truly fundamental and objective terminology must be welcomed.

Crenel Physics intends to be one such an initiative to 'restart physics from scratch'. So to speak with some castle walls, thereby first creating the virtual stronghold called 'Crenel Physics'. And initially there is *nothing* contained within these walls. And from this completely empty shell onwards, *no* property, no standard, no unit of measurement is accepted as a 'family member'... unless it truly represents something objective, natural, and new.

It thereby turned out, due to the intended 'grass roots' approach, that while underway a part of the aforementioned 'Planck units' have been more-or-less reconstructed. Was this a re-invention of the wheel? Yes!

But the reasoning behind it comes from a different angle: it is bottom up and therefore relatively easy to follow for a broader audience. Albeit, that in doing so a slightly different basis for 'normalization' popped up (relative to mainstream literature). Consequently the outcomes for conversion factors between metric units and the found *natural* units of measurement (see chapter 8) slightly differ from the aforementioned 'Planck units'.

> The followed authentic and 'cold eyes' reasoning is intentional: it is a key requirement for a restart from scratch. The aforementioned slight difference in 'normalization basis' is the price to pay. The 'authentic reasoning' should thereby not be confused with an arrogant attitude towards settled and highly respected physicists. It seems justified here, because it keeps the reasoning relatively easy to follow. And as said: although slightly deviating from the main stream, this book thereby is not unique in its outcome with respect to 'natural units'. Uniqueness may however lie in the logical extension of the reasoning thereafter, resulting in an interpretation of 'gravity' and a 'basic model'.

To avoid potential confusion, the terminology 'Planck units' will not be used anymore in relation to the here described theory. The similarity in outcomes will nevertheless be obvious, and actually confirms consistency with the mainstream.

Crenel Physics also is intended as one low-threshold communication platform for fundamental physics. It intends to further clear some skies to a potentially larger audience, and thereby nevertheless performs the complex task to give or develop a set of 'universally objective yardsticks'.

Ultimately –perhaps after many human generations- these type of efforts may lead to one ultimate goal: that newborns learn to base both their daily as well as their scientific thinking on 'natural units' alone. And that in doing so, they will speak the natural language of the entire universe in which they live.

There is a possibility to host a potential community of contributors on the internet at: www.crenelphysics.net. The reader is encouraged to check this site for contributions and developments.

2. 'Packages'.

This chapter:
- Illustrates through an example how Metric Physics units of measurement indeed gradually became scattered in the cause of history,
- Introduces the 'Package' as the first physical property in Crenel Physics.

Consider the historic development of the physical property *quantity*.

Physics is amongst others about 'matter', whatever *that* may be. And one of the first questions thereby is: *how much* matter… is subject to an experiment? The *quantity* of matter is one *key* physical property.

What is the *quantity* of a stone? Is nowadays in Metric Physics the procedure to numerically determine such a property (the quantity of this stone) fixed and settled? Perhaps not surprisingly, the answer to that question is: 'no'. That is why the property *quantity* can serve as a good example here. It illustrates the historically grown scatter and ambiguity in terminology.

Initially, the *quantity* of a stone was determined by the human effort that is required to lift it. Thereafter, and relatively speaking not too long ago, weighting scales were introduced. These express *quantity* as 'weight'. 'Weighting' is still widely used for practical purposes. However, the numerical result of 'weight' depends on *where on Earth* the measurement is done. The physicist Newton discovered that a weighting scale measures how strong a stone is pulled downward towards the Earth. Because the Earth was found to be rotating, it was recognized that the stone experiences -besides 'weight' caused by gravity- a centrifugal force which is highest at the equator, and zero on the north-pole (or south-pole). Therefore the numerical outcome of weighting the stone appears not equal at a global scale.

> Furthermore, because Earth itself experiences a centrifugal force, it has the shape of a slightly flattened sphere. At the poles of the Earth the stone would be closer to the Earth's center of gravity. This is another effect causing the weighting scale to show a higher value for 'weight' at the poles, relative to the equator. And a third force -impacting the outcome of 'weighting'- would be the weight of the air in which the stone is residing, causing an upward lift. This force is subject to local air density. Finally: in a spaceship orbiting the Earth, the stone would be found weightless…

Therefore, the 'weight' of a stone cannot be determined just anywhere. And *if* there is a resulting number, it does not uniquely specify 'quantity'. The deficiencies of the method gradually became clear as science developed.

As an improvement for 'weighting', the weight of the stone can be compared with some 'standard stone' that one would need to carry around. That's quite a burden. But nevertheless: today it still is common practice to carry 'standard weights' around for this purpose.

That's one room in Metric Physics, dealing with 'quantity'.

> From a physical viewpoint, determining 'weight' by *comparison* gives a *relative* unit of measurement. However, scientists realized that they needed *absolute* units of measurement. *Universal* physics is only possible through *absolute* units of measurement. Note: *absolute* units of measurement are not to be confused with the previously discussed and ultimately targeted *natural* units of measurement.
>
> The term 'absolute' requires, that one could give instructions by phone to someone -anywhere- on how to measure the unit in question, using local means only. One cannot send the required 'standard stone' through the telephone. Nor can one send an Earth or any other heavy object to produce the required gravity. Therefore, weight is *not* an absolute unit of measurement.

To cope with the aforementioned problems associated with the determination of 'weight', a new unit of measurement for 'quantity' was introduced: *mass* (symbol 'm'). The amount of *mass* is determined by applying a known force (symbol 'F') to the stone, and subsequently measuring the resulting acceleration (symbol: 'a') that this stone experiences as a consequence. The benefit of this concept is that it would not matter where on Earth -or in space- this measurement is performed: the outcome would always be the same. Therefore, at first sight the *mass* indeed seemed an *absolute unit of measurement* (provided, that one has absolute methods for quantifying force and acceleration).

The underlying formula is:

$$F = m.a \qquad (2.1)$$

In the metric system the force 'F' is expressed in Newton, the mass 'm' in kilograms, and the acceleration 'a' is in meter/second^2. Thus, if one applies a force of 1 Newton to a stone, one now needs a *stopwatch* and a *yardstick* to determine how long it takes before the stone accelerates from e.g. zero speed to a speed of 1 meter per second. Should this take 25 seconds, the mass of the stone would be 25 kilograms per equation (2.1).

That's the second room in Metric Physics, dealing with 'quantity'.

This method was good enough until about one hundred years ago. It then became clear that the quantity of the stone is *only* 25 kilograms if it is not moving relative to the observer. As soon as it has any relative speed -and it does not matter in which direction this speed is- the stone will appear to have *more* mass. It became clear that *mass* was depending on the velocity relative to the observer. That's quite a problem in Metric Physics because due to the nature of the measurement of mass, one *must* accelerate the stone to some extent. And one therefore must thus deal with the consequences of velocity, even while performing the measurement.

Thus, the concept of *relative mass* (m_{rel}) was introduced as refinement of the mass found at rest (m_0) relative to the observer. This *relative-to-the-observer* mass is calculated as:

$$m_{rel} = \frac{m_0}{\sqrt{1-\frac{v^2}{c^2}}} \qquad (2.2)$$

In which 'v' is the velocity of the stone relative to the observer, and 'c' is the velocity of light in an empty space (both velocities are in meters per second). As soon as the observer measures a relative velocity of an object, this formula for 'relative mass' is to be applied.

And therefore formula (2.1) is only valid if the *relative mass* per formula (2.2) is entered into the equation, thereby replacing the variable 'm'.

Furthermore, it was found that the *velocity of light* ('c' in equation 2.2) appears *not* to be a relative velocity. No matter how observers move around: through sections of vacuum (or empty space) each individual observer will always find and measure the *same* numerical value for 'c'. The latter is in conflict with day-to-day experiences with velocities. Evidence is however overwhelming. Physicists had to accept that the laws of physics are one thing, and day-to-day human experiences another. The difference is dramatic when high velocities are involved: according to equation (2.2) the relative mass of any object of substance would be infinitely large when the light speed would be assumed.

That's the third room in Metric Physics, dealing with 'quantity'.

Nevertheless, even without fully understanding the whereabouts of the above, and without fully understanding the 'constant result of light velocity as measured in vacuum while one is moving around', one might still argue that the above findings are not too complex to work with.

But the scientific developments did not stop here.

It became clear -beyond doubt- that *mass* (in kg) and *energy* (symbol 'E', expressed in 'J' which stands for Joules) are mutually convertible. A certain amount of mass represents a certain amount of energy, and vice versa. These conversions actually and verifiably take place in the world around us. The conversion formula is famous, and anybody will recognize the hand of Albert Einstein:

$$E = m.c^2 \qquad\qquad (2.3)$$

While arrived here, a small sidestep to *economics* seems appropriate to illustrate an important consequence for Crenel Physics:

> Assume the aforementioned stone is for sale. Europeans can buy it for € 500.00 and Americans can buy this very same stone for $ 700.00. Assuming the conversion rate between Euro's and Dollars is fixed (here at 1.4 $ per €), it really would not differ if payment is in Euros or in Dollars. No matter what currency is used: the seller can accept either the Dollars or the Euros (or any mixture). The conversion -back and forth- can be done through any bank. Having concluded

this, it than is reasonable and also logical to state that the stone actually has one -and no more than one- underlying *property*, namely its economic value, its *price*. This property *price* can be expressed in both Dollars and Euros, or in any other currency.

The above formula (2.3) describes an identical situation. That's because in this formula the speed of light 'c' is found to be a universal and absolute physical *constant*. The factor c^2 in this formula is therefore nothing but a *universally fixed* (fixed for all, under any circumstances) conversion rate between two units of measurement, namely *energy* at the left side of the equation, and *mass* at the right side.

The relevant question for supporting the targeted lean approach in Crenel Physics is: although one is -in practice in Metric Physics- using *two* different units of measurements (Joules and kilograms), how many *real* physical properties do these two units of measurement represent? There can only be one logical answer: 'mass' and 'energy' represent *one* -and no more than *one*- physical property…

And given the rigidly lean approach to follow, Crenel Physics cannot accept *both* units of measurement (Joules *and* kilograms). It is unfortunate that physicists did not know all of the above *before* the Joule and the kilogram were introduced as *apparently* separate properties.

Since this equality/exchangeability between Energy and Mass became known, the one -and single- underlying physical property for 'quantity', this *unit of measurement*, did not receive a single and unique *name* for common usage. Despite the aforementioned three rooms in Metric Physics, all dealing with 'quantity' and all using different units of measurement, the underlying commonality remained under the snow. And the human brain adapted to an *apparent* diversity. As the human brain likewise adapted to various currencies, each of which is expressing one and the same underlying property: the economic value.

Therefore this fundamental physical property, this single unit of measurement for *quantity*, will now receive a unique name in Crenel Physics:

> *Definition 1: 'Package'*
> *Package = the fundamental physical property (or unit of measurement) of a matter, expressing its quantity.*

According to formula (2.3), Packages can appear as either *energy* or *mass*. Both appearances are two different perceptions of the *same* physical property. And there are additional appearances of the Package. These will be discussed later.

This example illustrates how - in the cause of history - Metric Physics gradually became scattered: the properties *energy* and *mass* were maintained side by side, even though they *do* represent the same underlying physical property.

To further clarify the consequences of the followed approach, the above historic overview and discussion is as follows recaptured, reworded and enhanced from the Crenel Physics perspective:

> In physics one wants to determine the *quantity* of matter at hand.

> First a unit called 'weight' was used. The 'weight' was initially determined by lifting the stone manually, thereafter by using a weight scale. But the only 'absolute' method appeared, to determine 'mass'. 'Mass' is quantified by applying a *force* during some *time*, and measuring the resulting *acceleration* (thereby continuously correcting the mass for its relative actual speed to the observer).

> One apparently can use a *force*, a *stopwatch* and a *yardstick* to determine this 'mass' …

> From a Crenel Physics viewpoint this rings an alert bell:
> somewhere or somehow there must be a very fundamental *physical* relationship -not just the given *mathematical* formulas- between 'mass', and these three means that are used to determine it (a force, a stopwatch and a yardstick). How else would or could one use these means for the purpose of quantifying 'mass'? One should be alert to not overlook such almost 'snowed under' relationships in the upcoming search for *true* properties: these relationships are *not* obvious to be identified.

> To further illustrate that the existence of a physical relationship indeed is a requirement: if

one wants to measure the temperature in a bathtub there *must* be a physical relationship between what a thermometer is reading, and the temperature in the bathtub. Obviously there only is such a relationship if one physically sticks the thermometer in the bathtub.

So where is this physical relationship (this bathtub) between 'mass' and force, between 'mass' and stopwatch, and between 'mass' and yardstick?

In Metric Physics the related units of measurement are kilograms, Newton, seconds and meters respectively. But one cannot find a fundamental physical basis for as many individual units of measurements. There only appears to be a mathematical relationships, such as the formula F=m.a. These mathematical formulas indeed reflect the actual behaviors in nature where these units of measurement come together.

But nevertheless: the listed individual metric units of measurement cannot possibly represent as many underlying and *fundamental* individual physical properties. There must be some physical overlap.

And even after having used a force, a stopwatch and a yardstick to determine the value for 'mass' in kilograms, one *still* must be careful: the resulting numerical value is correct only, if the stone is not moving relative to the observer. Anybody who passes by will disagree with the numerical result. The moving traveler will argue that the mass of the stone appears *higher* to him. And again: this effect is *physical*, although it can be quantified through a mathematical expression (equation 2.2).

The example illustrates that in Metric Physics the system of physical properties is *not* reflecting the underlying fundamentals of physics. There is *no* one-on-one relationship between the properties in use (expressed in their associated units of measurement) and the fundamental physical properties, as a naïve by-passer may expect.

Metric Physics obviously is based on what once were *thought* to be true fundamental physical properties.

Planck's law enhances the above viewpoint even further. This is Planck's law:

$$E = h.\upsilon \qquad (2.4)$$

In which 'E' is again – as in equation (2.3) - the amount of energy contained by an object, expressed in J (Joules). The variable 'υ' represents the *frequency* of the object (the number of cycles per second). And the symbol 'h' (Planck's constant) represents a universal physical constant, which acts as a *conversion factor* between these two units of measurement (here *energy* and *frequency*).

Through Planck's law, the aforementioned scatter of terminology in Metric Physics is further enhanced as follows: neither the weight, nor the energy, nor the frequency of an object represents an exclusive physical property. Instead, all three are to be quantified by using the earlier defined property 'Packages'.

The above also illustrates that the scatter in Metric Physics goes straight to the fundamentals of physics. That fact alone justifies a reconstruction from the ground up.

3. 'Photons' (1) and 'Strength of Gravity'.

Until here, *one* fundamental physical property was defined in Crenel Physics: 'Packages' (see definition 1). It was discussed that Packages can manifest themselves (or: appear) as mass, or energy or frequency.

This chapter enhances the list of appearances of the Package even further, but first introduces the Photon.

Through relatively simple experiments, *Photons* do indeed appear in all three aforementioned forms. Therefore they will be subject to further analyses here. A Photon is often referred to as a 'light particle', although Photons are not necessarily in the range of visible light. Photons are electromagnetic particles that each individually reside somewhere on a very broad electromagnetic spectrum. On this electromagnetic spectrum, Metric Physics uses various compatible yardsticks: for example eV, frequency, wave length, mass, etcetera. The position on the electromagnetic spectrum –and thereby the numerical value of each yardstick- is *relative* to the observer: as soon as the observer changes his velocity relative to the Photon, its position on the electromagnetic spectrum changes.

Other than in Metric Physics, in Crenel Physics only one yardstick will be in use to pinpoint a Photon on the Electromagnetic spectrum: the Package.

In the past, the study of visible light caused a lot of debate between physicists: initially it was not clear that these aforementioned metric yardsticks on the electromagnetic spectrum all actually do represent *the same* fundamental physical property.

Nowadays however, experimental capabilities are more sophisticated. Individual Photons can even be isolated and detected as separate entities. Furthermore, for Photons (and also for 'light') one can quantify the Metric Physics units of measurement. Photons *do* appear as:

- Mass
  E.g.:
    o the path of a Photon (and 'light') is 'bended' if it passes the gravity field of a heavy star,
    o the tail of a comet (consisting of particles) is pushed away from the Sun through sunlight impact,
    o a fragile – almost frictionless - 'windmill' (e.g. mounted in a vacuum bulb) starts turning by shining light on it, demonstrating the impact of impulse power: Photons indeed are considered as mass containing particles.
- Energy
  E.g. heat is generated if a Photon (and 'light') is absorbed at a black surface. The capturing of Photons may cause heat, secondary radiation, an electron voltaic effect (as in solar panels).
- Frequency
  one measures wavelength, e.g. by diffraction after impinging of Photons on a screen with a small opening.

Because in Crenel Physics all three manifestations quantify *one and the same* fundamental property, namely 'Packages', there is a built-in presumption for a fundamental connection between the three associated measurable properties as well. These manifestations cannot possibly be seen as non-related to each other anymore, for the single reason that they all will lead to *exactly* the same numerical result (when the quantity of a Photon is expressed in 'Packages').

In Metric Physics, the relations between the above appearances are mathematically expressed through the equations (2.3) and (2.4) in the previous chapter. The concept of the Package in Crenel Physics is entirely based on these equations.

On top of the aforementioned appearances, the newly introduced fundamental physical property called 'Packages' has additional forms or manifestations through which it can be revealed. More in general: it even is thinkable that these manifestations do not necessarily have to appear in 'our world'.

'Gravity' is another –easy to find and additional- manifestation of the property 'Packages'. This is from a Crenel Physics perspective, as will be discussed below.

Once an object (like a Photon) contains a certain non-zero number of Packages, it thereby inherently appears to receive a physical feature which here will be called 'Strength of Gravity'. Only objects containing some non-zero number of 'Packages' appear to have some of this 'Strength of Gravity'.

When two objects - that *both* contain at least some Packages- are within each other's mutual reach, they *always* appear to apply a mutual attract force to each other according to the 'law of gravity'. Thereby, the strength of a resulting 'gravity attract force' appears to be proportional with the number of 'Packages'. Or: if one consider two objects, *both* should contain at least some 'Packages' in order for a *mutual* and *equal* 'gravity attract force' to exist. This force then inescapably becomes apparent. And there seems no way to shield it. And should *any* of the two objects contain more 'Packages', this mutual and equal attract force inescapably increases proportionally between *both* objects. The rate at which this 'gravity attract force' increases – proportional with the combined number of Packages - appears to be a fundamental universal physical constant. In Metric Physics this rate is referred to as the 'gravitational constant'.

Based on the observation that this 'Strength of Gravity' property is one-on-one related to the 'number of Packages at hand', it is concluded here that it does *not* represent a new property in Crenel Physics:

> The number of 'Packages' *is* the 'Strength of Gravity'.

Whether one looks at the 'number of Packages' or at the 'Strength of Gravity' of an object: one is looking at *one and the same* property. They apparently cannot be separated from each other.

In Metric Physics the concept 'gravity' is presented as an independent feature, having impact on 'mass'. In Crenel

Physics however, through the Package there is an integrated (and inclusive) representation: objects that contain Packages *are* attracted to each other through the manifestation 'Strength of Gravity'.

There is a potential for confusion between the term 'gravity' in the classical sense, and the newly introduced term 'Strength of Gravity' (which is actually: Packages). To avoid such confusion, in Crenel Physics terminology the implications of classical 'gravity' are worded as follows:

> If an object1 contains a certain number of X Packages, its 'Strength of Gravity' is equal to X. And if an object2 contains Y 'Packages', its 'Strength of Gravity' is Y. The 'gravity attract force' between object1 and object2 is than proportional with the product of X and Y (or XY). And it will *also* be proportional with the *gravitational constant*.

> This *gravitational constant* will be discussed later in more detail.

Should an object have the physical property of *zero* 'Packages', this would imply that it contains zero mass, *and* zero energy, *and* a frequency of zero, *and* zero 'Strength of Gravity'. Thus, there would be nothing to detect in Crenel Physics (given its current level of development), since any detector would require at least some exchange of *any*thing. And this object would have no things (nothing) to exchange in the Package arena. And so far, in Crenel Physics that's the only arena.

The 'gravity attract force' strictly and exclusively appears between *pairs* of objects.

The *possibility of interaction* between entitics – here through a 'gravity attract force' – is a feature that almost unnoticeable sneaked into Crenel Physics. The objective of Crenel Physics is however to follow a rigidly lean approach. The acceptance that a 'gravity attract force' may appear between a pair of objects implies – with some reservations at this point – that *separate* entities can reside. This requires –in accordance with the human perception- that separate entities individually can reside somewhere, sometime. And that they can interact through a force. Neither the *somewhere*, nor the *sometime*, nor the *force* have been shaped and introduced into Crenel Physics yet: they will be discussed later. Based on know how from Metric Physics it nevertheless is possible to – preliminary – further explore some features of this 'gravity attract force' appearance of the Package.

As stated before, the 'gravity attract force' only appears between a pair of objects, when each object contains at least some Packages. The relationship between a specific pair of objects is presumed -based on Metric Physics findings- to be *exclusive*: it is neither shared with other objects (that may be around), nor can it be shielded e.g. by other objects. Thus, if one considers a cluster of e.g. four objects that are within each other's reach, each individual object will form three exclusive pairs (with the remaining three objects). And thus it presumably executes (and experiences) three completely independent and exclusive gravitational relationships. When one of these three remaining objects would be removed, this would have *no impact at all* on the remaining two relationships. At first sight and by intuition this presumed complete independency of relationships between pairs may seem irrelevant and rather trivial. But at the bottom line (and thus in Crenel Physics) the relevance of this presumed complete independency of relationships is paramount.

The presumed absolute independency of (gravity) relations between pairs of objects causes (or explains) that in Metric Physics the *local* observations will predictably -through the theory of relativity- differ from *remote* observations. In Metric Physics the two objects that together form an independent relationship, do so within their mutually shared frame of reference. This shared frame is by definition a one-dimensional (spatial) frame: in Metric Physics two points in space define one (and no more but one) *spatial* dimension. Remote observers may however travel around this one dimensional spatial frame, or may see it move around or rotate from their perspective. Remote observers will therefore consequently 'see' relative masses (and thereby different applicable forces of gravity), different distances (and thereby – again – different applicable forces of gravity), and different timeframes. All of this is pending their relative motion.

But that is not all: in Metric Physics it is also recognised that the 'relativity of motion' is in turn a relative concept by itself. One could thus have both the 'pair of objects' *and* 'the observer' move around in a broader third spatial frame of reference. And seen from this third frame, one can *still* predict (quantify) the observations that the aforementioned –moving- observer will do from his position. These relative calculations can -at closer look and at bottom line- only be found consistent because all three parties involved (the studied pair of objects, the remote observer, and the scientist using a broader frame of reference around these both) share *one* particular universal physical property: the velocity of light. This velocity appears equal to all, regardless their relative movements. This equality is embedded in the formulas of the previous chapter. And because this velocity of light is equal to all, it causes observed masses, distances, forces and energies (energy is also force multiplied by distance) to shift in a consistent and predictable manner.

In Crenel Physics, none of these relative units of measurement (used in Metric Physics) have been introduced yet. In fact, the entire concept of 'space' and 'movement' needs further investigation in order to decide how it can be reflected. In Crenel Physics, there is no relativity yet. At this point it is even very questionable whether there will be relative units of measurement… One would rather expect that in Crenel Physics the theory of relativity is not a theory, but an integrated fact. In that case there should be no relative properties or units of measurement. This expectation will be subject to later verification.

At this point it is concluded that the introduced 'Strength of Gravity' does *not* represent a new feature in 'Crenel Physics'. It is fully represented by 'the number of Packages'.

Crenel Physics can therefore -until here- be summarised as follows:

Photons are entities that contain Packages.

The velocity of Photons (or: 'light') is accepted as a universal constant (although 'velocity' as a concept has not yet been introduced into Crenel Physics).

The number of *Packages* expresses a physical property that can appear as:
- Energy,
- Mass,
- Frequency,
- Strength of gravity,
- And perhaps as other appearances (such as 'information content', 'number of degrees of freedom', etcetera).

Interactions are exclusive relationships between pairs of entities.

4. Crenels.

This chapter:
- Discusses 'time' and 'distance'
- Introduces the 'Crenel' as physical property in Crenel Physics.

An object can only be detected if it contains at least *some* (non-zero) number of Packages. It then is one of the *appearances* of the 'Package' that will be detected, and not the Package itself.

Metric Physics demands another –second- requirement that must be met to make detection possible: an object must also exist for at least some period of *time*. This is not always as easy as it may seem at first sight. Particles – in their monitored appearance - may indeed come and go. And some particles have very short lifetimes (or formulated more accurately: the type of *appearance* that actually happens to be monitored has a very short lifetime). Physicists sometimes have to deal with very narrow time windows for detection.

What exactly is 'time' or a 'time window'?

Should 'time' be allowed into Crenel Physics as the second physical property, or perhaps as a *variable*? The latter seems a 'contradictio interminus': 'time' would than vary in 'time'. 'Time' would than vary within itself...

To assess the potential meaning of 'time' in Crenel Physics, it first will be investigated what 'time' really stands for in Metric Physics. Here, it is defined as a physical *variable*: 'time' is measured in seconds or fractions thereof. And this variable 'time' is found in many physical formulas, therein represented by the symbol 't'.

Where does the numerical value for this variable 't' come from, prior to being entered into these formulas?

Perhaps the first method for quantifying 'time' was to count how many times the Sun was visible... and disappeared again. Days were followed by nights, and thus the number of repeating sequences was counted. Although the underlying physical process initially was not understood (only later it became apparent that the Earth was a globe rotating around its axis), this worked. One even found a repeating pattern in the daily maximum elevation of the Sun above the horizon, and thus started to count one year for approximately every 365 days. Here, the underlying physical process was the Earth orbiting the Sun. And again this was not understood initially, but that did not really matter to the end users.

Later the pendulum was invented. It turned out that a weight hanging on a rope would swing to and fro at a constant frequency, provided that both gravity and the length of the rope are constant. Counting the number of swings thus became an additional measure for time. One advantage of this pendulum was that one could manipulate the length of the rope, such that some elementary physical rules became apparent per Huygen's equation:

$$\text{Swing time } = 2.\pi. \sqrt{\frac{\text{rope length}}{\text{gravitational constant}}} \qquad (4.1)$$

And using these rules, individual pendulums could be fine-tuned between them to swing at equal pace. This created a new shared time standard. One advantage was, that the frequency was high relative to the day and night rhythm on the Earth, and that therefore pendulums could be used to measure shorter periods of time, like hours, minutes and seconds.

Nowadays, one counts a certain oscillation in a Cesium-133 crystal. It has been universally agreed upon, that after 9,192,631,770 of these oscillations *one second* of time has passed. That is a very accurate definition. Even very short time windows can now be counted: as short as 0.000,000,000,1 seconds.

This is so accurate, that if one would take two equal Cesium-133 crystals, it is *not* to be taken for granted that one would see them run at exactly the same pace. The laws of relativity in Metric Physics state for example, that a clock appears to run *slower* if it is moving away from the observer. If the observer however keeps such a clock around his wrist, he would never see this particular one run slower. This particular one would simply tell him how fast he is growing older, and that there is no escape from that. It is the moving clocks *around him* that may appear to run slower relative to his clock.

To investigate this, assume that a friend sits beside the observer in deep space (where nothing else is around), and that the observer gives him a clock like the one he keeps on his wrist. They check times, and make sure that their clocks are synchronised. As long as they stay closely together, both clocks will stay in tune and synchronised.

Now, the observer asks his friend to make a space walk, away from him. While he walks away, the observer can presumably see this other watch. And as long as the second watch is walked away, it appears to run *slower*. This is all according to the laws of relativity in Metric Physics. There would be no difference in whether the friend walks away from the observer, or the observer walks away from his friend. In both cases the observer would see the other watch run slightly slower relative to his own. There is complete exchangeability and symmetry.

Let's assume that after a while both stop their relative motion. From that moment onwards, both watches will appear to run at the same pace again. But the continuous increment of the earlier lagging behind - that the observer saw accumulate on his friend's clock while he was moving away - will not disappear: this accumulated difference will now remain constant. This implies that the accumulated lag time that the observer sees on his friends clock (relative to his own) is *the* measure to quantify how far they actually are separated: one does not need an extra instrument for determining distances. Comparing the available clocks will do.

Both would now express the distance in (lag) *time* units, which would be *seconds* –probably small fractions thereof- in Metric Physics.

The amount of the found lag-time is equal to the time it would take electromagnetic particles (or: 'light', or 'Photons) to travel from the friend to the observer, or the other way around.

> This experiment in free space contains *the* most relevant key fact in physics: on the basis that –at any time- there can be only *one* physical distance between the observer and his friend, while this distance is measured in travelling time of electromagnetic particles, it *must* be so that these electromagnetic particles travel at equal speed relative to *both* perspectives (here: the observer and his friend). The physical fact that 'light velocity (in vacuum) is a universal constant for all' thus can alternatively be reformulated as: 'between two entities there can be only *one* –and no more than one- physical distance at any moment in time'. The second formulation will raise few eyebrows and even sounds trivial. Light velocity as a universal –non relative- constant is much more difficult to accept. Yet, both formulations express the same physical fact.

> Note that this experiment does not imply that both the observer and his friend will judge the length of this distance –when expressed in meters- to be equal under *all* circumstances. This will be discussed later.

At first sight the above may sound somewhat theoretical and of little practical value in daily life. But many of us witnessed this effect for the first time when man landed on the Moon: the limited speed of radio signals obviously slowed the conversation with Houston. If an astronaut on the Moon would tell when he saw that it was 12:00:00 hours sharp at Greenwich (according to his observations), down on Earth one would hear his check call about 2.6 seconds too late. And that's not because these astronauts were sloppy men. That's because it takes 1.3 seconds for light to travel from Greenwich to the Moon, and another 1.3 seconds for the radio signals (which are also Photons or electromagnetic particles) to reach Earth. Because electromagnetic particles travel at about 300.000 km/s, the distance between the astronaut and Earth would not only be (approximately) 400,000 km, but *also* (approximately) 1.3 seconds.

Nowadays, using lag time measurements for distance has entered our daily lives: it is what the global positioning system (GPS) is based upon. The distance from the observer to – at least four - individual satellites is very accurately *measured* in lag time units relative to the local clock. These four individual and different lag times are determined, and subsequently converted into individual distances to as many satellites. Given the well known locations of these satellites, the location of the observer on Earth (or near the Earth) can then be calculated. In doing so, at bottom line the lag time units are converted into geographical coordinates e.g. on electronic maps of the Earth surface, plus an altitude relative to the local sea-level. Several secondary corrections for the speed of electromagnetic particles -en route- are applied, but these are not relevant here.

It is only because -nowadays- one can measure time windows very accurately (using e.g. a Cesium-133 crystal) that one can indeed determine distances in ranges that are of the daily usage. The aforementioned minimum detectable time window of 0.000,000,000,1 seconds is the lag time that would correspond with a distance of approximately 0.03 meters.

The opening question of this chapter was, whether 'time' should be introduced as the 2nd -next to Packages- unit of measurement in Crenel Physics. This must now be evaluated from the above perspective: the underlying physical property is *not* just 'time', but it is -from a physical perspective- also completely synonym with 'distance'.

It is fact that in daily life and perception, humans nevertheless make a major difference between measuring 'time' and measuring 'distance'. However, from a physical point of view there is no difference whatsoever. If one looks at his hand, one does *not* just look at something which is half a meter away: one is also looking at history. Both ways of expressing that the hand is not where the eye is, are equivalent. Actually, one would be looking approximately $0.5/3x10^8$ = 0.000,000,013 seconds back in time (note: the factor $3x10^8$ is the approximate speed of light in m/s). Or: one is looking at where the hand *used* to be 0.000,000,013 seconds ago.

When distances are small, humans prefer the unit 'meter' for distance. In astronomy however distances are large and the 'light year' is thus a more practical unit of distance measurement. A star can be 500 (light) years away. And one's hand is only 0.000,000,013 seconds away. The wide spread in range makes it understandable that from childhood onwards humans programmed *two* different units of measurement (meters and seconds) into their brains. One simply applies two different yardsticks for one and the same physical property.

One question then is, whether humans also have *two separate sense organs* that inform about 'time' and 'distance' respectively… and separately. It is clear that humans *perceive* both as completely separate units of measurement. These perceptions are however subjective. It indeed is questionable whether other life forms (intelligent or not) have a likewise perception of 'time' and 'distance'.

As discussed in chapter 2, in Metric Physics there was no name found for what lies underneath the perceptions called mass, energy, frequency, or strength of gravity. Only the human perceptions have been named. To fill this gap, in Crenel Physics the 'Package' was introduced. Here –in a likewise situation- a unambiguous name for the fundamental physical yardstick to measure the perceptions for distance and/or time is needed in Crenel Physics.

Prior to introducing this new name, there is yet another fundamental consideration. The question 'how does one actually quantify time in Metric Physics?' has only been slightly addressed so far, and now it becomes relevant. At the bottom line one is counting a 'number of repeating occurrences'. Whether one is counting days (revolutions of the Earth around its axis), years (revolutions of the Earth around the Sun), swings of a pendulum, or oscillations in a Cesium-133 crystal: all this boils down to one and the same principle.

The most basic 'thing' that could repeat its status would be the 'binary function', as discussed in chapter 1. It has only *two* statuses. And it would alternate between these. No simpler form of 'change' is thinkable. The binary shape was associated with crenellation. This lead to the naming: 'Crenel Physics'. Inspired by this, the unit of measurement 'Crenel' is introduced here:

> *Definition 2: 'Crenel'*
> *Crenel = the unit of measurement (or physical property) to quantify time or distance between 'objects'.*

Thus, in Crenel Physics, there is neither 'time', nor 'distance': there only is the 'Crenel'.

> 'Time' and 'distances' are considered *appearances* of the Crenel. This is consistent with considering 'mass', 'energy', 'frequency' and 'Strength of Gravity' (and others) to be *appearances* of the Package.

Any of these aforementioned appearances may –or may not- be perceived. It solely and entirely depends on the sense organs (or instruments) that are in use, whether (or not) a certain appearance will be perceived. If an appearance is not perceived, it still is there and ready for perception. If one closes the eyes (or: shuts down a sensor), this does not mean that light disappeared…

Based on the above discussion, it now is appropriate to re-raise the initial point (here slightly fine-tuned) from the beginning of this chapter and originating from Metric Physics, namely that:

An 'object's monitored appearance' must exist *at least for some period of time*, to make detection possible.

At first sight this requirement may have looked trivial, but in light of the above discussion: what exactly does this mean? It seems impossible to grasp the true physical meaning of the concept of 'time' or 'distance'. The bottom line is that in practice one would count 'Crenel' (or some other property) that *periodically repeats itself*. Based on today's knowledge one can do no more - and no less - than just that.

However: *what* exactly is a periodical repetition? Whatever may happen in the cause of a 'Crenel' seems impossible to understand. Perhaps in one 'Crenel' a lot of something called 'time' passes by, while in the next 'Crenel' much less 'time' passes by. There seems to be no way to determine or understand what is actually happening underneath while a 'Crenel' is counted. Does each oscillation in a crystal really represent an equal amount of elapsing 'time'?

It so seems, that for practical reasons physicists simply reversed the reasoning: here the unit of 'time' (the second) *is* represented by 9,192,631,770 oscillations in a Cesium-133 atom. While in reality it is not clear what this 'time' really means. Of course physicists are aware of the very fundamental question raised here, but nevertheless: when they use the symbol 't' -for 'time'- in formulas, they 'de facto' refer to a number of repeating occurrences (oscillations) in *matter*, and *not* to time itself.

Until today this has been a successful approach, because there is a tight physical relationship between the 'velocity of light in vacuum' at the one side, and 'oscillations in atoms/crystals' at the other side.

When a velocity is measured, the required 'time' basis is based on these 'oscillations in atoms/crystals'. And in doing so –by choice– the resulting light velocity appears to be universally constant. Only recently the required *distance* basis was redefined in Metric Physics: one meter is now defined as the distance that light travels in vacuum during 9,192,631,770 oscillations in a Cesium-133 atom. It is through this latest definition of 'distance' that the light velocity does not just *appear* to be constant, but is actually *presumed* to be constant, and thereby serves as (one of) *the* absolute and universal reference(s) for units of measurement.

However, *if* one would understand what 'time' really stands for, e.g. the radioactive decay of a single radioactive atom could be predictable. Based on the current frame of reference one now must consider such radioactive decay of a single atom - within some given timeframe - as an 'at random' occurrence. Some physicists believe that knowledge will come where such coincidence can be taken out of the equation. In light of the above discussion it would require a breakthrough in the understanding of 'time', delivering a new and better yardstick that would eliminate e.g. the aforementioned coincidence in radioactive decay.

To complete this fundamental thought about elapsing time, consider the –weird- alternative to use the Geiger counter ticks near some radioactive standard source somewhere in universe as the basis for any time measurement: assume that after each tick, one time unit has elapsed. This alternative choice for time unit definition would cause atom oscillations to appear as an at random occurrence and the value of the light velocity would be all over the place… Such a choice would have the consequence that all current units of measurement would seem to float around at random. However, the radioactive decay of the aforementioned radioactive standard source would thereby be entirely predictable in 'time'.

In Crenel Physics one will have to accept –at least for now– that the real meaning of 'time' (and thereby distance) is *not* understood.

And as discussed earlier, one consequently has to keep in mind that if one wants to communicate with other life forms, the human appearances 'time' and 'distance' may *not* be mutually shared… In fact: even other living species on Earth (like horses) may not perceive these appearances as humans do.

Coming back to the initial requirement in this chapter, that an objects (or better: the monitored *appearance* thereof) must exist at least some period of time in our world in order to be able to detect it: that same requirement can now be reworded as:

A 'monitored appearance' must be associated with something that has changed -at least some of its- status.

Or, to rephrase this conclusion:

> *'Without any change of status (in some system anywhere), one will **not** be able to detect anything'.*

This conclusion –without its given context– looks strange and abstract. However, within the above context one will recognize that 'some status change' really is the only basis for detecting the *appearance* of 'time'.

It now seems appropriate to associate the above with another point that was made earlier:

> If one wants to determine how many Packages an object contains, one can use: a force, a stopwatch and a yardstick. With the force applied, one will accelerate the object under investigation, and the rate of acceleration allows the calculation of the number of Packages, using formula (2.1).

The point made was, that *because* this is a valid procedure to follow, there *must* be a physical relationship between mass/energy/frequency at the one side, and force/time/distance at the other side. There *must* be a relationship between the property one is measuring, and the tools one is using to measure it.

Such a relationship is not obvious if one evaluates the units of measurement in Metric Physics. But such a relationship should -or *must*- become obvious while defining new units in Crenel Physics. That is: as long as one sticks to the strategy to follow the fundamentals of physics, and nothing else.

Now that it has been settled that:

- Mass, energy, frequency, strength of gravity (and perhaps other appearances) are quantified using one and the same physical yardstick, named Packages,
  and that
- Time and distance (and perhaps other appearances) are quantified using one and the same physical yardstick, named Crenel,

the search for such a relationship can start.

Such a relationship can indeed be found in Metric Physics.

Part of it is already given in Planck's law: here, energy is related to frequency, where energy is expressed in Joules, and frequency in –reciprocal- seconds (*reciprocal* seconds is equal to seconds^{-1}).

But there is another relation in Metric Physics, according to the theory of relativity. It so appears that if one clock stays under the Eiffel tower (or any other tower), while a friend's clock resides on top, both clocks will *not* exactly run at the same pace anymore, despite the fact that the mutual distance is kept constant. The theory of relativity says that under the influence of a stronger *gravity field* at the Earth's surface, the observer will notice his clock run *slower* relative to the friend's clock up there on the tower, where the Earth's gravity field is slightly weaker. And the friend's clock will run *faster* than the observer's. Both will mutually agree upon the difference of pace.

More in general, Metric Physics says that the more gravity one experiences, the slower the clock runs relative to a clock that is in deep space, far away from any mass.

Since very accurate clocks are available nowadays, this can actually be verified.

> Prior to having these very accurate clocks, this same phenomenon has already been found through the analyses of the light frequencies that various atoms or gas ions emit. The light frequencies originating from atoms at the Sun's surface appear to be slower than those which are generated by the same type of atoms located on Earth. That's because the Sun is much heavier than the Earth, and thus has a much stronger gravity field. Of course these frequencies would also shift if the Sun moves towards -or away from- Earth (the 'Doppler effect'), but both effects are different and can be identified separately.

So there *is* a relation: if there are a lot of Packages near a clock, according to a remote observer this clock appears to run slower relative to the absence of any 'Packages'. Although this is well documented in Metric Physics, it will be explored again –later- from the perspective of Crenel Physics (see chapter 15).

Note: the mission remains to base Crenel Physics on 'true' physical properties only, by keeping the system of units of measurement as lean as possible. This implies that physical relations known from Metric Physics are to be taken into account after careful analyses only. For example: the aforementioned clock slowdown explained by the presence of a 'gravity field' would not hold in Crenel Physics: here there is no gravity field as such. It is clear that such a clock slowdown takes place, but the reasoning behind it must come through some alternative reasoning. This will be discussed in chapter 15.

5. Virtual appearances, π, and Photons (2).

Chapter 2 introduced the physical property Packages. In Crenel Physics the term Packages reflects both the name for a *fundamental physical property*, as well as to its *unit of measurement*: the property 'Package' is expressed in 'Packages'. And likewise, the 'Crenel' is expressed in 'Crenels'.

Note the difference in a *fundamental* unit of measurement, and a *natural* unit of measurement. The Crenel and Package have been defined as *fundamental* units of measurement. They are however not *natural* units of measurement because there is an –earlier discussed- anticipated relationship between the Crenel and the Package. Based on this relationship, one may assume an even more fundamental unit underneath the Package and the Crenel. Therefore, the Package and the Crenel are not qualified as *natural* units (see chapter 12).

It also was discussed that Packages and Crenels each have different *appearances*. Packages can appear as 'mass', 'energy', 'frequency', 'Strength of Gravity'…and perhaps more. Crenels can appear as 'time', 'distance'… and perhaps more.

Within Crenel Physics, 'appearances' are defined as follows:

> *'Appearances':*
> *are signals that an observer may sense with his sensors.*

The selection of sensor(s) that an observer is actually using may depend on availability, on the nature of the experiment, etc… This selection sets -and limits- the particular appearances that one may become aware of. The appearances that are *not* registered during a particular experiment are nevertheless valid and in place. To use a metaphor: light does not disappear by closing the eyes, or by having insensitive eyes.

Any of the aforementioned examples of appearances is expressed in either Packages or Crenels. Thereby it does not matter which of the appearances one may happen to monitor: when expressed in these units of Crenel Physics, the numerical outcome will be the same.

It is typical for science in general that most of its units of measurement are –historically- based on *appearances* rather than on fundamental properties that lie underneath. Thus, the appearance 'mass' is expressed in 'kilogram', the appearance 'energy' is expressed in Joules, etcetera. In Metric Physics these units of measurement are named *fundamental*. This is however not correct because there is a more fundamental layer underneath. In Crenel Physics, the aforementioned units of measurement are therefore named *appearances* of the underlying fundamental physical properties: the Package and the Crenel.

The Crenel –being the shared unit of measurement of two well known appearances 'time' and 'distance'- is hard to envision by the human brain. However, the impact of its *appearances* on human thinking can be found almost *anywhere* and *anytime*… Even the here used terms *anywhere* and *anytime* are the result of this impact. A quick review of physical formulas in Metric Physics illustrates that the symbol 't' (for 'time' in seconds) as well as the symbol 'd' (for 'distance' in meters) are frequently used. And thereby these are considered to be completely separate and independent parameters. Metric Physics is indeed adapted to appearances, to this *human perception of the world*.

Slightly more than 100 years ago physicists managed to look beyond these human perceptions (is there a human who really *understands* the physical analogy between distance and time?). In doing so they unintentionally pulled science away from the human comfort zone. It was acknowledged that *appearances* were not the leading indicators anymore. Appearances remained however anchor points to the human daily thinking, and thereby to science. This may explain why in Metric Physics e.g. the 'theory of relativity' was perceived as a *refinement* of science, rather than as an integrated part.

'Time' and 'distance' are nowadays the truly adopted son and daughter of one single parent: here named the 'Crenel'. The found relationship between 'time' and 'distance' was a major scientific breakthrough: one second converts to 'c' meters. Thereby, 'c' is the conversion rate and has a universal fixed value of approximately 300,000,000 meters per second: it is the speed of light as measured in vacuum. Despite the simplicity of this conversion rule this type of relationships between appearances make nature feel unnatural.

So many fundamental relationships between various appearances were identified, that reversing the reasoning also became an option. In science, in addition to the traditional top-down search for *relationships between various appearances*, a bottom-up search for *potential appearances* is going on as well. Thus, appearances that never had been 'seen' –that never appeared through detectors- were predicted… and found. Sometimes dedicated new detectors needed to be developed first. This bottom-up search often takes *fundamental* physical properties as starting point.

To appreciate the deep footprint of appearances in the human thinking, one question to investigate is: how do humans actually manage to perceive *multiple* 'appearances' based on just *one* fundamental physical property underneath? And furthermore: are these indeed just 'appearances' (as suggested here)… or are these perhaps for 'real' (whatever 'real' may be)?

Chapter 13 discusses the appearance 'color'. It illustrates the human creativity in perceiving 'color appearances'. It explains how humans manage to live in a colorful world that requires three independent parameters to model it, while underneath there is only *one* single physical parameter: the Package containment of individual Photons.

The question '*why* do humans perceive an appearance like colors?' is outside the scope of this book. Here, it can only be concluded that it is a fact. And that this became so through the evolution of living species. Not only through the evolution that shaped mankind, but also through evolution of the individual person since his birth (at birth, colors are not yet consciously perceived).

The human perception of 'distance' can be evaluated in a likewise manner as the perception of colors. As discussed earlier, 'distance' is just one appearance of the fundamental physical property 'Crenel'. One should not assume that *other life forms* perceive the same appearance of 'distance' as humans do... perhaps this appearance is even exclusive to -or even typical for- humans.

'Distance' is -in the human brain- associated with a perception of some 3-dimensional spatial fixture, or frame of reference. Within such a frame the appearance of 'distance' is just one of the *two* relevant components. The second is 'angles'. These are required to express -relative- *directions*. Humans need both a 'distance' perception *and* an 'angles' perception to construct their 3-dimensional spatial fixture.

'Space' will be discussed in more depth later. Here, the concept of 'angles' is addressed first.

Crenel Physics requires a yardstick for 'angles', to indicate their size. To find such a yardstick, one can start by imagining a 2-dimensional 'plane'. Because humans are adapted to the appearance called 'distance' *and* have a perception for 'angles', one can easily envision such a 'plane'. In doing so one has to realize that a plane is nothing but a *mathematical* appearance. There is no reason to assume that such a 'plane' could or would indeed *physically* exist. A 'plane' has a thickness of 0, and therefore it cannot contain anything substantial. A 'plane' is useful to structure the human thinking, and to communicate concepts or ideas. The language of communication is called *mathematics*. A 'plane' is nothing but an appearance… in the human *imagination*. It therefore is a *virtual* appearance.

Although a 1-dimensional 'line' and a 2-dimensional 'plane' are *abstract models,* one can nevertheless buy a meter of rope, or a square meter of land. So there *is* practical relevance to these virtual appearances.

In the aforementioned virtual 'plane' one can now envision a mathematical concept that is called a 'circle'. Within this plane a 'circle' is defined as: the collection of all points within the plane that lie at some arbitrarily chosen -but equal- *distance* from one single arbitrarily chosen *point* in this plane. In Crenel Physics this arbitrarily chosen distance would be expressed in Crenel, in Metric Physics it would be expressed in meters. In both cases it is referred to as being the *radius* of a circle.

It was found that -if the plane is 'flat'- the circumference of this circle will always equal exactly $2.\pi$ times the arbitrarily chosen radius. Here, the symbol 'π' stands for a dimensionless and universal constant. Its numerical value is approximately 3.1416. The exact numerical value of 'π' cannot be exactly expressed in a decimal numerical system: the number of digits has no end. The exact value is the outcome of a *procedure*: divide the length of a circle's circumference by its diameter. It is the numerical outcome of this procedure that received an exclusive name (and symbol) in science: 'π'.

The mathematical constant 'π' will be allowed into Crenel Physics:

'π' is the ratio between the diameter of a circle, and the length of the circumference of that circle.

And -using this mathematical model of a 'plane' and a 'circle'- one now can introduce an *absolute* measure for 'angles' or 'rotation'. To achieve this, the 'radial' is introduced as the *absolute mathematical* unit of measurement for 'angles'. Traveling one full circle represents 2.π of these 'radials' (or: 'number of radiuses') of angle rotation. Travelling just a part of the circle represents a proportional fraction of 2.π 'radials' of rotation. An angle's value does not have to be limited to one full rotation: an angle can e.g. be 25.π radials, which represents the equivalent of 12.5 full rotations or revolutions.

One now can give instructions -a procedure- by phone to create a rotation or angle of e.g. 250.345 'radials'. That possibility makes this unit of measurement (the 'radial') *absolute*. This fact (and thereby the absoluteness of 'π') is as important to science as the invention of the wheel is to automotive industry.

One must nevertheless be careful while applying the above mathematical procedure in physics. Care should be taken when 'distance' comes into the picture. And it did while defining the radius of the aforementioned 'circle'. From a mathematical viewpoint the 'distance' between two points on a rope is only equal to the length of the rope -between these two points- when the rope is 'straight'. The definition of the radius of the aforementioned circle requires a full understanding of what *distance* means. Or: what 'straightness' (of a rope) really means.

To verify 'straightness', one should take the imaginary rope and ensure that –firstly- it is entirely embedded in the aforementioned 'plane' *and* –secondly- that within this plane the rope would indeed appear to be 'straight' and is not zigzagging or curved. But when these two requirements have been positively confirmed, this raises a third point: how can one verify that this 'plane' is flat? If for example an area of this plane has a hilly shape, the rope still may not be 'straight' even though this may appear so within the plane itself. To verify this 'flatness', one would have to put the 2-dimensional 'plane' into a 3-dimensional 'space'. And indeed one can now verify that within this 3-dimensional space the 'plane' appears to be 'flat' (and therefore the rope will indeed be 'straight').

Until here the procedure to verify straightness is easy to envision. The next step is more difficult: how does one verify that the 3-dimensional space –used for the verification- is not 'curved'? Until not too long ago that question would have raised eyebrows. Nowadays however it *is* fact in Metric Physics that '3-dimensional spaces' can indeed be 'curved', namely under the influence of a gravity field. Therefore, for verification one would have to put this 3-dimensional 'space' into a 4-dimensional 'space' to verify that it indeed is not curved. But who ensures that in turn this 4-dimensional 'space' is 'straight'? One could verify this by putting this 4-dimensional 'space' into a 5-dimensional space, and so on, and so on.

From a *mathematical* viewpoint there really is no argument to stop verifying straightness in increasingly higher spatial dimensions. And thus the straightness of a rope can 'de facto' never be verified in this manner… and therefore mathematics alone cannot answer the original question: how is the required 'equal distance' between all points on a circle towards its centre ensured? Without this insurance it will not be possible to use the radial for angle measurement.

> Therefore, from a mathematical viewpoint one can neither verify the aforementioned circle, nor can one apply the procedure to quantify π as a universal constant, nor the 'radial' as an absolute angle.

But *physics* makes it possible to follow a more natural approach: go back to the original question by *re-defining* 'distance', e.g. in accordance with Crenel Physics (see the previous chapter). Here, the described procedure to quantify 'distance' was as follows:

1. Synchronize two clocks at some point 1 on the rope,
2. Have a friend walk with one clock to a second point on the rope, and
3. Measure the resulting lag-time between both clocks.

The found lag-time is expressed in Crenels. It is equal to the time that it takes *light* (or *any* Photon, or electromagnetic particle) to move from the watch at point 2, towards the watch at point 1.

The rope is 'straight' if the measured lag-time (expressed in Crenels) is equal to the length of the rope (also expressed in Crenels).

In Crenel Physics, a 3-dimensional space has not yet been introduced. At this point only the matter of 'straightness' has been addressed, such that π can be quantified. And per the above, it is in effect the path of some free traveling Photon that is –at the bottom line- taken as *the* definition of 'straightness'. The reasoning is thus reversed (relative to Metric Physics).

In Crenel Physics not only the *velocity* of a Photon is taken as a standard for reference, but also the *path* of a Photon is a standard for reference: this path is *the* definition of 'straight'.

> Note that the path of light can indeed be curved or bended (e.g. through lenses). In chapter 16 the directional changes of light beams are discussed: such curving or bending is *not* equivalent to curving an individual Photon's path. Individual Photons *do* travel straight.

By using individual Photons as reference for 'straight path' determination, there are two physical considerations to further investigate here. These considerations both are based on the fact that in Metric Physics the path of Photons through a 'Metric Physics frame of reference' may be curved by a gravity field, whereas in Crenel Physics this path *is* the definition of 'straight'.

The first consideration is that heavier Photons (those with a high frequency) would expectedly experience a different gravity force relative to lighter Photons (those with a lower frequency). The concern is, that through some gravity field the heavy Photons than would potentially follow another path than lighter Photons. That -of course- would cause ambiguity in the aforementioned straight path definition in Crenel Physics, and thereby inconsistency between Metric Physics and Crenel Physics. The question than is, whether in Metric Physics all Photons –regardless their mass (or frequency) - would indeed follow the *same* route through a frame of reference, when gravity is around.

To further analyze this first consideration, the force of a potential gravity field can be split into two directional components:

1. A gravity force component *in parallel* of the direction of propagation of the Photon (or into the exact opposite direction).

   This component would be applicable e.g. to a Photon that is emitted by the sun in a direction which is vertical relative to the sun's surface. This force component would -by definition of the Photon- have no impact on the Photon's velocity: the 'velocity of a Photon' would – trivially- still remain the 'velocity of a Photon', which is *the* one and only synonym to the 'velocity of light' and set as standard for reference….
   Instead, this force component will reduce (or

increase) the energy containment by the Photon. According to Metric Physics this force would reduce (or increase) e.g. the Photon's contained appearance of 'mass', or 'frequency', or... etc.

2. A gravity force component that is *perpendicular* to the direction of propagation of the Photon.

   This component would be applicable e.g. to a Photon that is *passing* a heavy object. Such a perpendicular force component does not impact the Photon's forward velocity, but –according to a Metric Physics spatial frame of reference- instead it results in curving its path.
   The more Packages a Photon contains (or: the heavier the Photon, or: the higher its frequency), the stronger the gravitational pull will be, and vice versa.

   Because the strength of the gravitational force, as well as the contained Packages of the Photon, are *both* equally proportional, the actual acceleration (and resulting curving of the path) in a gravity field will therefore *not* depend on the individual Photon's frequency (or other appearance).
   Thus, the curving will be equal to all Photons, regardless their Package containment.

Thus, according to both Metric Physics *and* Crenel Physics, through a gravity field all Photons (heavy or light) would follow an equal and unambiguous path at equal velocity.

Thereby -in Metric Physics- the yardstick for 'distance' (the meter) is not an absolute unit of measurement: it is *relative* as it e.g. depends on gravity fields. Thus, in Metric Physics –when looked at from a remote position- the 3-dimensional space appears 'curved' under the influence of a gravity field. And the Photon follows its route through this 'curved' space. If from a remote position one would take its own yardstick and physically travel to this 'curved' area, the local yard still appears equal to the original length of the yardstick. *Locally* one therefore would not be able to confirm any curving. The conclusion is that the 'curving of space' can only be observed remotely, and *not* locally. Relativity is only observed when local units of measurement are applied to remote areas. Or –more in general- when there is some distance between the yardstick(s) and the observation(s).

In the spatial fixture of Crenel Physics however, not only the path of the Photon remains straight (this path is the *definition* of straight). The spatial fixture also remains straight, even when evaluated from a remote position. That's because in Crenel Physics all distances are measured in Crenel, which in turn are based on Photon path and velocity properties. In Crenel physics Photon paths are not curved (by definition), and Photon velocity is constant, and therefore the spatial frame of reference is not curved. Space therefore is Euclidian.

The above equality of the followed paths by a series of different Photons is based on the assumption, that the Photon is passing the gravity field of a very heavy object (e.g. a planet). In the reasoning it was assumed that it's flying by has no impact on the gravity field that it is flying through. There is however some gravity interaction between this very heavy object and the Photon. But the path of the very heavy object itself will only theoretically and marginally be affected by the reaction forces of just one Photon during its flyby.

Things would be different -and this leads to the *second* consideration- if one envisions a situation where a potential and disturbing gravity field (making the Photon's path curved in Metric Physics) is caused by a very light object at very close range. Obviously, such a close flyby (near miss) of the Photon would indeed notably change the path of the other object as well, due to a reaction force. For this situation one can envision e.g. two Photons that pass each other at very close range. In such a scenario, according to Metric Physics one may expect that the heavier Photon's path is less bended –relative to the path of the lighter Photon- due to the gravitational interaction.

> From a 3-dimensional Metric Physics viewpoint, this scenario would cause *scattering* of the path of the Photon. This would however be limited to a *directional* scattering, as the forward velocity of the Photon remains unchanged. Due to this directional scattering, it would take a Photon more time to cover the distance between a point 'A' and a point 'B'. And a next Photon, commencing the trip, would experience another

directional scattering pattern and therefore follow another spatial path. In this scenario, very heavy Photons would on average experience less directional scattering relative to lighter Photons. Thus, on average heavy Photons would cover the distance between 'A' and 'B' faster than light Photons.

Because in Crenel Physics the Photon paths are used to determine the distance between two points, one must make sure that these Photons do not impact the gravity field of any other objects in their closeness. This requires that the Photons one is using as 'distance measuring devices' cannot invoke gravity forces to other objects. This consequently requires that these Photons -used as distance measuring devices in non empty spaces- contain a hypothetical number of *zero* Packages (or mass, or frequency).

Photons have been allowed to enter 'Crenel Physics'. Here is the definition:

> *Definition 3: 'Photon'.*
> *A 'Photon' is a container for 'Packages' travelling at the 'velocity of Photons'.*
>
> *This –trivial- velocity is also called the 'velocity of light'. It is relative to any observer, no matter where this observer is residing or how fast he is moving.*
>
> *The true special feature of Photons is that they **all** travel at equal velocity, regardless directions and regardless the number of Packages that the*

individual Photon may contain. From a spatial viewpoint Photons cannot and will not overtake other Photons. They can only cross roads (including head-on collisions).

It should be noted that it has been a *choice, a degree of freedom,* to declare the velocity of light being a universal constant. Consequently, the appearance 'time' became a relative unit of measurement in Metric Physics. Technically one could as well have taken 'time' as the universal unit of reference. For example: the frequency of a Cesium crystal could be assumed universally equal anywhere, from any perspective. Had this alternative been selected as *the* standard, the velocity of light 'c' would have been a relative unit of measurement. At the bottom line, whichever option is in fact chosen has no *fundamental* relevance: it would not change physics itself. It would only impact the units of measurement in which processes are described. Therefore, in Crenel Physics the same choice is made as in Metric Physics: the velocity of light is set as *the* universal reference. In doing so, 'time' has become a *slave*.

At this point, and for the purpose of completeness, the 'zero-Photon' is introduced in Crenel Physics as a special type:

> *A 'zero-Photon' is:*
> *a hypothetical Photon which contains zero Packages (or with a frequency of zero, or a mass of zero, or an infinite wavelength, or… etc.).*

Such 'zero-Photon' would be an ideal device for determining distances, given the earlier consideration that it may not impact the gravity field through which it is propagating.

However, with its hypothetical infinite wavelength, from a spatial perspective it is questionable where in space the zero-Photon would reside at any moment in time. It would not be possible to locate it, and hence it would also not be possible to follow its path. From a spatial viewpoint (where wavelength is a key appearance of the 'distance' type) a zero-Photon would be anywhere and nowhere at the same time. This completely and entirely disqualifies 'zero-Photons' as distance measuring devices.

This spatial ambiguity of the zero-Photon raises the question how accurate the spatial grid in Crenel Physics can be defined, and how the position of a 'normal' Package containing Photon therein can be determined. A typical Photon does have a finite wavelength and the question than is how accurate its position could be determined, and also how far its spatial influence would reach.

In Metric Physics, there is the uncertainty principle (formulated by Heisenberg). According to this principle, at first sight in Crenel Physics the location of a Photon would be totally uncertain because its velocity is exactly known (in fact: this velocity is used as a standard). In Crenel Physics this same uncertainty principle is –of course- also valid, but it will need to be reformulated based on a different system of fundamental units of measurement. The Heisenberg uncertainty principle relates to *appearances* (velocity and spatial position of a particle), and not necessarily to the underlying more fundamental units of measurement as used in Crenel Physics.

Heisenberg uncertainty principle in Crenel Physics is reflected in the limited accuracy at which a spatial frame of reference can be defined. In Crenel Physics, spatial space is not a mathematical –virtual- coordinate system: it is based on Photon properties and behavior. And pending the wavelength of a Photon, there is uncertainty with regards to the exact spatial whereabouts.

With the Crenel Physics definition for 'distance', the question whether a Photon did indeed follow a 'straight' path (or not) has become irrelevant. In fact, the whole concept of 'straight' has become irrelevant: instead one needs to follow the path of individual Photons. And the length of this path is the Crenel Physics definition of 'distance', which in turn is expressed in Crenels. It will be the only definition that is applied in Crenel Physics:

> *'Distance' is along the path that Photons would take. It is measured in Crenel.*

> Note - for comparison - that in Metric Physics the 'meter' nowadays is defined as the distance that is covered by light in vacuum during a timeframe of 1/299,792,458 seconds, while in turn the 'second' is defined as the amount of time that elapses during 9,192,631,770 oscillations in a Cesium-133 crystal…where the pace of these oscillations is a relative unit of measurement that depends on relative velocity *and* on local gravity fields.

Within Metric Physics this indeed closed the circle in recognizing the full implications of taking the velocity of Photons (in vacuum) as reference for monitoring units.

The Crenel Physics toolbox now contains all elements to construct a 'human' 3-dimensional spatial frame of reference in which an imaginable experiment can take place. For this frame one would need the following tools from the toolbox:

- 3 instances of the 'distance appearance' of the 'Crenel', and
- The mathematical concept of 'angles', measured in radials where $2.\pi$ radials cover a full circle.

However, as color seeing is totally appearance based, within this frame of reference one also is residing in *nothing but appearances* of only *one* fundamental physical unit: the 'Crenel'… one is living in appearances of a 'fundamental world'. And whatever the 'fundamental world' may *be*, is irrelevant in this practice… the 'fundamental world' *appears*.

That is bad news: there is a difference between an *appearance* and an *existence* of a process. And thereby things are even worse: the appearances that actually will be perceived are entirely depending on the nature of the sensors one is using. If one closes the eyes, the *appearance* of the world changes, but the actual properties of this world do not change a bit…

Humans –using their natural sensors- would be quite familiar with the aforementioned appearance of a 3-dimensional space. And obviously it is reasonable to base practical science on this windowed view.

With this spatial frame of reference in place, there still is nothing *in* it. It is completely intangible. This is where the Package comes into sight. In Crenel Physics, Packages represent tangible *content*, while at the same time Packages are associated with an appearance called 'frequency' (and thereby allow time measurements).

Thus, Packages and Crenels are the Yin and Yang of Crenel Physics (although here they are not opposites or counter poles to each other… but -as will be discussed later- they are *reciprocal* to each other). To construct a world, one will need *both*.

6. Spaces, Fixtures and Virtual Spots.

To the human brain, the 'motion' of an object appears to take place in a 3-dimensional *spatial* space. In Crenel Physics this space is nothing but an *appearance*. It is based on three *distance appearances* of the Crenel, each pointing in a different *direction*. From the Crenel Physics viewpoint there is no argument against envisioning a spatial space as such, as long as it is clear that this is just *one* of the potential appearances. One may replace one or more *distance* appearance of the Crenel with *time* appearances. Such replacement then would describe this *same* 3-dimensional space. At bottom line one is using the one and only underlying Crenel anyway.

This chapter analyses spatial spaces and introduces the *Fixture* and the *Virtual Spot*.

There are various mathematical methods to define coordinates a 3-dimensional spatial space.

A commonly used method in Metric Physics is based on an X, Y and Z-axis that are perpendicular to each other (a 'Cartesian' frame of reference). Any point in space can then be pinpointed, using the three associated coordinates. These are referred to as 'rectangular coordinates' or 'Cartesian coordinates'. This method is typically used for situations where no symmetry between the coordinates is obvious.

Another method uses two *angles* and one *radius* (or distance), measured from a central point where the three aforementioned axis cross. Again, any point in a 3-

dimensional space can be located by using these 3 parameters. This method is also based on a frame of reference set by an X, Y and Z-axis. Usually, the two angles are named 'φ' and 'θ', and the radius 'r'. These are called 'spherical coordinates' or 'polar coordinates'.

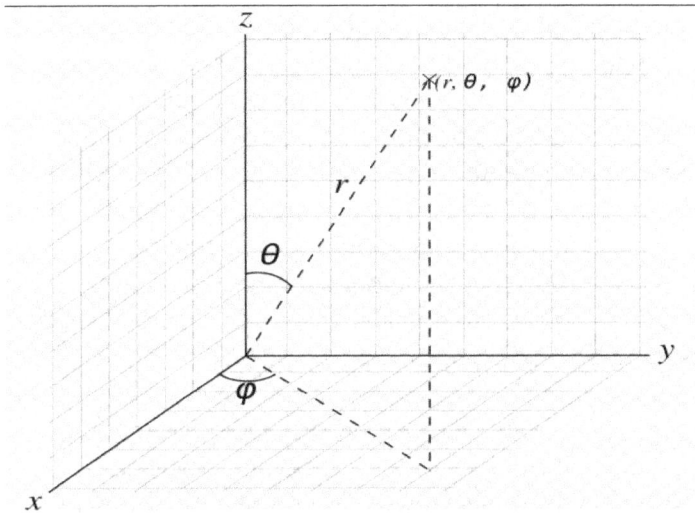

Figure 6.1: spherical coordinates or polar coordinates.

Spherical coordinates are typically used for situations where complete symmetry is expected in all directions. For example to describe a gravity field around a spherical object.

A third method of describing a 3-dimensional space is through 'cylindrical coordinates'. This method is typically used for situations where cylindrical symmetry is expected, for example the gravity field around a pair of two separate objects. And again, this method is based on a frame of reference set by an X, Y and Z-axis.

Conceptually, all methods fully overlap each other. The various methods do not rule out each other: any method can be applied.

The difference is the complexity of the mathematical equations that will come forth from a certain analyses. If an 'unhandy' method is selected e.g. for a 3-dimensional spatial analyses, mathematical equations can quickly become complex. For example: expressing the surface of a sphere with radius 10 is simple in spherical coordinates. The equation is: 'r =10'. That same spherical surface, expressed in rectangular coordinates, would be: $x^2+y^2+z^2=100$. That is a more complex equation. And it would be difficult to describe this same spherical surface using cylindrical coordinates.

Spherical coordinates are well fitted to express some fundamental rules in physics: the single distance coordinate (or dimension) 'r' is completely synonym with the measurement of 'lag time' between a central point of reference, and any remote location anywhere in space.

> *Using spherical coordinates, the 'lag time relative to the observer in the centre' is not to be seen as some fourth (time) dimension in a 3-dimensional space. In spherical coordinates this **is** one of these dimensions, namely the radius.*

While rectangular coordinates typically contain three coordinates that *all three* are of the *distance* appearance, the spherical coordinates contains *two* different types, each having its own appearance and unit of measurement:

- One *type of coordinates* is of the distance type: the radius 'r'.
  Using spherical coordinates in a 3-dimensional space, there is only one coordinate of this type. In Metric Physics the radius coordinate is measured in e.g. meters, or light years. In Crenel Physics it is measured in Crenels. And these Crenels can have –at least- two appearances: the 'Crenel$_D$' or the 'Crenel$_T$', where the indexes 'D' and 'T' refer to *distance* and *time* respectively.

- The second type of coordinates is the 'angles', of which there are two instances in a 3-dimensional space using spherical coordinates: the horizontal angle φ and the vertical angle θ. These are angles that are measured relative to a *reference direction*.
  Both in Metric Physics as well as Crenel Physics, angles are measured in 'radials'.
  In an empty space (or: Euclidean space), 'radials' are an *absolute* unit of measurement. Here, one full circle equals 2π of these 'radials', where 'π' is a universal constant.

For describing a 1-dimensional space one only needs *one* single reference. Thus, there is only *one* parameter (or: coordinate) –relative to that reference- that matters. By intuition humans would easily envision this single parameter as a *distance* from the reference point. And because the degrees of freedom within a 1-dimensional space are limited (the single parameter can only increase or decrease in its value), the envisioning of a 'line' may come up first. The human intuition often leads to envisioning this 'line' embedded in a 3-dimensional space. From a 3-dimensional spatial perspective this is however *not* correct: the single coordinate must –from this perspective- *not* be seen as a single point on some imaginary line, but instead as an entire sphere around the single point of reference within this space. Only the latter envisioning acknowledges all the extra features that a 3-dimensional space offers above the features of a 1-dimensional space.

There is a very essential difference between *wrongly* envisioning a 1-dimensional space as a series of points on a line embedded in a 3-dimensional space, and in *correctly* envisioning this 1-dimensional space as e.g. a series of concentric spheres embedded in this 3-dimensional space. The envisioning of a 'line' is wrong, because in order to fix a 'line' one would need *two* points of reference rather than one. Therefore, a 'line' has fewer degrees of freedom than a series of concentric circles.

> Should one take a paintbrush in a 1-dimensional spatial space, and paint a selected specific uniquely identified location red, from a 2-dimensional spatial viewpoint an entire circle would be painted red, and from a 3-dimensional spatial viewpoint an entire sphere would be painted red.

Furthermore, a 1-dimensional space (as any space) does not need to be *spatial*.

The single coordinate may as well be an angle (relative to a reference direction), expressed in radials. There would be no maximum to the value of an angle: $2.\pi$ radials (which equal approximately 6.28 radials) represent one full revolution, but an angle may as well be 250 radials. This would represent a series of full revolutions, plus some remaining angle. Or it may be a negative -250 radials: the same angle, but measured in the opposite direction.

Or one may as well consider the appearance 'electromagnetic spectrum' as a 1-dimensional space. This clearly is not a *spatial* appearance. Here, the single coordinate sets the (imaginary) 'position' of a Photon on an 'electromagnetic spectrum'. The position on this spectrum is then associated with a single dot, that represents the contained number of Packages, and thereby is associated with a fixed value for its appearances: an *energy level*, or a *mass*, or a *frequency*, etcetera.

Any 1-dimensional space can be envisioned to be embedded in a higher dimensional space of the same nature. This would also apply to the aforementioned examples 'angle' and 'electromagnetic spectrum': these would then be embedded in a higher dimensional 'angle space' or 'energy space' respectively. From a 2-dimensional perspective a pinpointed single coordinate would represent the equivalent of a circle, from a 3-dimensional perspective a sphere. Although this enhancement cannot be spatially *envisioned*, it can be *imagined*. From the Crenel Physics viewpoint there is no argument to treat *spatial* space differently relative to other types of 'spaces'.

In Crenel Physics, a 1-dimensional system therefore is anything that requires one parameter (and no more than one) to be pinpointed. It does not need to be *spatial*.

In Metric Physics the same approach is followed, but the spatial element is here often -by human intuition- perceived as factual and special, rather than being just one possible instance of a 1-dimensional space. The wording 'space' promotes the human envisioning of a *spatial* appearance.

One objective of Crenel Physics is to avoid ambiguous terminology. Because there is no fundamental or conceptual difference between a 1-dimensional *spatial* system and any other 1-dimensional system, a new and more general –less insinuating- term is introduced here for 'space'. This new term is: the *Fixture*. A Fixture in Crenel Physics is the same as a 'space' is in Metric Physics, but the term is not suggesting a relation with a potential spatial appearance of 'space'.

> A **Fixture** is an appearance with a number of independent parameters or variables. The number of independent variables is referred to as 'dimensions':
> - An appearance with 0 independent variables is referred to as a 0-dimensional Fixture (or: dimensionless Fixture). It has no reference point or reference direction or reference value, and therefore it has universal validity.

- *An appearance with 1 independent variable is referred to as a 1-dimensional Fixture.*
  It has one point of reference (or one reference direction, or one reference value).
- *An appearance with 2 variables is referred to as a 2-dimensional Fixture. It has two points of reference (or reference directions, or reference value, or any combination thereof).*
- *And so on.*

Here are some examples of Fixtures:

- 'π' is a 0-dimensional Fixture.
  It needs no reference at all: 'π' has its absolute and *natural* value.
- The Package containment of a Photon is a 1-dimensional Fixture.
- The temperature in a bathtub is a 1-dimensional Fixture (the variable is the reading of one thermometer).
- The *density* of an object is a 4-dimensional Fixture (the variables are spatial volume –which is a 3-dimensional Fixture-, and mass),
- The velocity of an object is a 2-dimensional Fixture (the variables are time and distance).

To become more familiar with this terminology, the law of gravity will now be described. A gravity field has a spatial shape, and it needs to be quantified at any point within that shape.

For the *shaping* a gravity field around a single object, a 1-dimensional Fixture will do. The concept 'direction' is irrelevant for this shape, and therefore is *not* a variable. One would envision spatial spherical coordinates as per figure (6.1), and position the object in the central reference point. The single variable would be 'r', to be expressed in Crenel. One should stay alert here: the human brain –trained in spatial envisioning- has indeed a tendency thereby to envision the *distance* appearance of the Crenel in this case, while the *time* appearance would be as valid. Both appearances are valid, and it depends on the type of sensors one is using, which one(s) will become apparent to the observer.

For *quantifying* the law of gravity, another superimposed 1-dimensional Fixture is required: the number of Packages that the object contains. In Crenel Physics the only object that can contain Packages and has been introduced so far, is the Photon. Thus to quantify the 'Strength of Gravity' of a Photon, one needs to determine its location on the electromagnetic spectrum (see chapter 13). The higher the number of contained Packages within the object, the higher its 'Strength of Gravity' will be.

Thus, to shape *and* quantify a physical fact such as the 'Strength of Gravity', one needs to superimpose *two* 1-dimensional Fixtures: one Fixture specifies the *shape* around the point of reference (expressed in Crenels), the other quantifies the *strength* by pinpointing where this point of reference resides on the Electromagnetic spectrum (expressed in Packages).

In this example, one now must mathematically connect these *two* separate 1-dimensional Fixtures to produce a new feature: 'Strength of Gravity'. According to the law of gravity, the 'Strength of Gravity' is the outcome of the 'Package Fixture', divided by the 'Crenel Fixture'.

Because one needs this mathematical combination of *two* completely independent 1-dimensional Fixtures for producing the 'Strength of Gravity' (associated with a single object), the shaping and quantifying of gravity requires 'de facto' a 2-dimensional Fixture.

The next step is to determine the 'Gravity Attract Force' between *two* separate objects. It is only when a second object is introduced that a concept like 'direction' (of Force) comes into the picture.

This 'Gravity Attract Force' is found by mathematically multiplying their respective 2-dimensional Fixtures for 'Strength's of Gravities'. This results in a 4-dimensional Fixture.

Even though the mutual distance is physically shared between both objects -and thereby shared by the underlying 2-dimensional Fixtures- the end result is nevertheless still a 4-dimensional Fixture. The analogy behind this is that e.g. a square meter (m^2) is indeed to be seen as a 2-dimensional feature, and *not* as two isolated instances of a 1-dimensional feature.

> In Metric Physics this spatial envisioning is quite intuitive, and thereby potentially biasing the thinking: if the unit of measurement 'm^2' is encountered, one easily envisions a 'square

meter'. If however the unit of measurement 's^2' is encountered (e.g. the acceleration of an object is expressed in m/s^2, in which the 's^2' is embedded), a likewise envisioning by the human brain is not available: no person can envision a 'square second'. Both are nevertheless 2-dimensional Fixtures, and both are based on the Crenel as the yardstick. There is no fundamental reason for the perceived difference: this difference is a *human* factor and thus a human reality. Other life forms however may have developed different realities.

When dealing with combinations of various Fixtures, there are some mathematical rules. Apples are not equal to pears: one cannot add them up (or subtract them). But in physics there apparently is no difficulty to *divide* apples by pears, or to multiply apples with pears, thereby creating new features (like 'gravity' in the above example). This can be rephrased into three more general mathematical rules that apply to Fixtures:

1. Any combination of Fixtures through multiplication or division is allowed, thereby creating more complex Fixtures.

   See the aforementioned example of determining the 'Gravity Attract Force', resulting in a 4-dimensional Fixture.

   Another example is *velocity*: this is an appearance where one Crenel Fixture (distance) is divided by another Crenel Fixture (time), which *mathematically* results in a dimensionless number. But that dimensionless outcome would not make the *Fixture* in which *velocity* resides 0-dimensional: this Fixture still is 2-dimensional. In Crenel Physics this is taken into account by differentiating between the *mathematical* approach (where Crenels in the nominator are equaled out with Crenels in the denominator, resulting in a dimensionless outcome, in a 0-dimensional result) and the Crenel Physics approach, where both units of measurement are kept in place: the *mathematical* result may be 0-dimensional here, but the physical Fixture remains 2-dimensional.

From a spatial viewpoint there would still be a directional component associated with velocity. In Metric Physics -where the mathematical approach is usually followed- the directional component is maintained through introduction of a *unit-vector* into the equation: this is a vector with a length of 1, which points into the direction of the velocity. In Crenel Physics, the directional information associated with velocity is *appearance* information. In Crenel Physics, the strategy to deal with this is that no dimensions or coordinates are eliminated through mathematical rules: units of measurement in the nominator cannot simply be equaled out by the same units of measurement in the denominator. This would lead to loss of dimension (and thereby physical information).

2. (Combinations of) Fixtures can only be added or subtracted if each term has the same dimension.

3. Equations -where at bottom line Fixtures are set equal to each other- are only valid if the Fixtures are of the same dimension.

These three rules are consistent with Metric Physics and with mathematical rules, even though mathematical rules would allow simplification of terms (reduction of dimensions): this mathematically *permissible* procedure is not a mandatory procedure, and is *never* to be applied in Crenel Physics.

These rules hereby are introduced into Crenel Physics. Here, these rules will expectedly cause less drama relative to Metric Physics because there are only two units of measurement available: the Package and the Crenel.

It seems logical that apples indeed cannot be added up with pears (per rule 2). The question then is, whether it indeed is allowed to *multiply* apples with pears (per rule 1). At the bottom line one is multiplying quantities of various appearances, for example:

3 apples x 4 pears = 12 apple.pears

Obviously from the above equation, the mathematical operation of multiplication (or division) results 'de facto' in nothing more than another *notation* of the same fact: the right side of the above equation is at bottom line nothing but another notation of the left side. It therefore cannot be associated with a true physical operation. From the Crenel Physics perspective, multiplication and/or division are not to be seen as *operations*: they are no more but alternative notations for the same feature.

> *The execution of multiplication and/or division (per the above rule 1) just leads to semantics for describing terms.*

> *The only physical operations (so far) in Crenel Physics are <u>addition</u> and <u>subtraction</u>. Like: 'A' + 'B' = 'C', where 'C' is indeed something new as a result of the operation. Here, through the operation, 'C' is something more than just 'A', and something more than just 'B'.*

This is not just an academic consideration. As was discussed in chapter 1, *computers* are devices that also can only execute numerical additions or subtractions (and move the results around in memory). Apparently, in Crenel Physics any process –no matter how complex- is to be seen as the ultimate outcome of an algorithm of additions and/or subtractions.

With the Fixture now introduced, there is no reason to ban the human 'spatial envisioning'. As long as one keeps in mind that one thereby is evaluating no more but just *one* specific appearance of a more fundamental reality that lies underneath: the Crenel. The 'time' appearance of the Crenel would do as well, and is also as present. The only reason that it is not perceived as such is that the human 'time' sensor is not suitable for spatial purposes. This alternative would therefore be harder –if not impossible- to envision. But nevertheless: it is there as well.

In Crenel Physics, it is now unambiguous to use the term 'space' for a *spatial* appearance exclusively. In cases where the fundamental parameters underneath are discussed, the term Fixture will be used. Thereby ambiguity is avoided.

From the aforementioned 1-dimensional *space* onwards, a further extrapolation was already addressed. For defining a 2-dimensional space one would require *two* points of reference (or alternatively reference directions). In a spatial view of a 2-dimensional Fixture there would be *two* coordinates that uniquely specify a location. E.g. the (one) distance from one of the two reference points, plus the (one) angle relative to the connection line between the two reference points. *Or* e.g. the two angles relative to the two reference points on their connection line (one angle measured relative to point 1, the other angle relative to point 2).

> The flexibility to either opt for 'distances' or for 'angles' being the spatial coordinates, requires that *relationships* between Crenels (for 'distances') and Radials (for 'angles') exist.

> In *mathematics* there are such relationships: for example Pythagoras's theorem, or the given that in a flat plane the three angles of a triangle add up to π radials. The question than is, if and when these mathematical rules and theorems are valid in Crenel Physics. It was already discussed that in Crenel Physics space is Euclidian. Therefore, in

Crenel Physics these mathematical rules apply. In Metric Physics space may be curved, and mathematical rules must be used with caution.

As discussed, from a 3-dimensional perspective, the aforementioned 1-dimensional *space* can only be correctly envisioned as a collection of spherical surfaces around some central point of reference.

And likewise, from a 3-dimensional spatial perspective the aforementioned 2-dimensional space needs to be envisioned as a collection of an infinite number of planes that -as a palmetto- share the connection line between the two reference points. Note: consistent with the earlier reasoning, a 2-dimensional space is indeed *not* to be envisioned as just one plane within a 3-dimensional space.

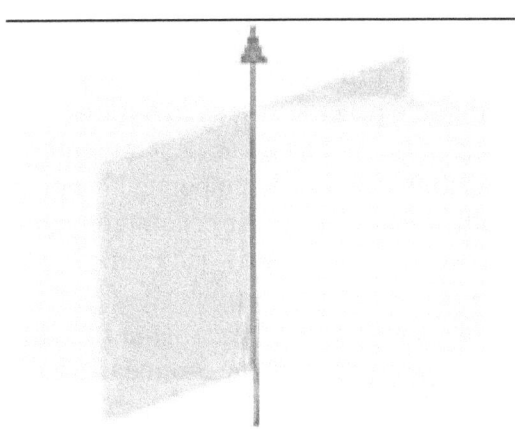

Figure 6.2: a 2-dimensional space from a 3-dimensional perspective shapes a palmetto of planes.

The above figure (6.2) represents a 2-dimensional space, seen from a 3-dimensional perspective. There is complete cylindrical symmetry (only 3 planes are shown for clarity, where in reality this 2-dimensional space consists of a palmetto of an infinite number of planes). Given a numerical value -for both the two coordinates in the 2-dimensional space- within each of these planes only *one* point is uniquely identified. From a 3-dimensional perspective however all identified points in the palmetto together form a *full circle* around the shared connection line between the two points of reference.

Therefore, if one would paint a uniquely specified point in a 2-dimensional space in a red color, from a 3-dimensional viewpoint one would see a red circle in space, around the connection line between the two reference points of the 2-dimensional space.

This demands some further reflection: the aforementioned red circle penetrates each individual plane in the palmetto at *two* points... one penetration point is the aforementioned uniquely pinpointed location within each individual plane. The other penetration point in the plane is however *not* pinpointed –and painted red- *within* that plane itself. Nevertheless, from a 3-dimensional point of view that particular spatial spot on the plane *meets the requirements*. In the 3-dimensional space this spot *would* be colored red.

This second point would correspond to the pinpointed location in one of the other infinite number of plane options: namely in that particular plane that is rotated half a turn relative to the plane under consideration. It is 'virtually' selected.

Within the 2-dimensional space this 'virtual location' can be found by mathematically *mirroring* the actually pinpointed location relative to the frame of reference of this space.

This can be summarized as follows:

> From a 3-dimensional spatial perspective, any pinpointed spot in a 2-dimensional space (a plane) is inherently associated with a pinpointed circle. And thereby this pinpointed spot is implicitly associated with a virtual location within that 2-dimensional space. From a 3-dimensional perspective this virtual location also meets the requirements of the coordinates.

Likewise, the aforementioned 1-dimensional space can be reviewed, from both a 2-dimensional *and* from a 3-dimensional perspective. This leads to the following more general conclusion:

> From a 2-dimensional and from a 3-dimensional perspective, any pinpointed location in a 1-dimensional space is inherently associated with a set of concentric circles or a set of concentric spheres respectively, and thereby with one unique *Virtual Spot* within that 1-dimensional space. A *Virtual Spot* that –from a higher dimensional perspective- also meets the requirements as are specified by the one coordinate, associated with the 1-dimensional space.

What would happen –in this spatial appearance of the Crenel- if one would paint a single point in a 1- or 2-dimensional space red? As discussed, from a 3-dimensional viewpoint an entire sphere or circle would thereby be colored red. This question then focuses on the *speed* at which this coloring would appear. The answer is, that the entire circle or sphere would be red colored in one single moment of time: *instantaneously*. A concept like *propagation* does not seem applicable here, because in fact one is only painting a single spot: that is inherently an instantaneous event.

And that implies that from within a 1- or 2-dimensional world one can conceptually communicate or interact at infinite speed –thus faster than light- from the perspective of a 3-dimensional world. Just by coloring one single dot red or blue (so to speak), in higher dimensions entire circles or spheres are colored red or blue instantaneously. This could be used as a kind of Morse code for communication. This communication could however *not* be detected from within the 1- or 2-dimensional world itself: here, *only one* selected dot is painted red or blue. Furthermore, this instantaneous communication is not possible between any two points in space: it is only possible between a selected spot, and it's associated and unique *Virtual Spot*.

The 'coloring' of the circle or sphere is a metaphor for an *intangible* operation: imaginary circles or spheres cannot be painted. Nevertheless, other intangible forms of communications or interactions are thinkable along this mechanism. *Intangible* means within this context that nothing substantial is involved in the operation.

In Metric Physics there are various experiments that suggest (or: demonstrate) *immediate* interaction between specific –related- points that are spatially separated. And furthermore, nature thereby reveals symmetry in many – if not all- situations. The apparent 'cross space interaction at infinite speed' is puzzling from a 3-dimensional spatial perspective. The above concepts are at the basis of such phenomena:

> By extrapolation, infinite speed interaction or communication within a 3-dimensional space would require a transfer of the aforementioned findings into a 4th spatial dimension: a 'dot' in a 3-dimensional space would than instantaneously shape a 'circle' in a 4-dimensional space. This would by implication result in an instantaneous *virtual* selection of a mirrored 'dot' within the original 3-dimensional space, *without* the direct possibility to notice this from the 3-dimensional perspective. Intangible interaction at infinite speed -between both spots- is thereby however facilitated in concept. And this communication would be limited between one (or: any) particular location in space and its associated *Virtual Spot*.

The following table summarizes this concept, thereby also allowing extrapolating the results into an undetermined number of higher dimensions:

SEE- DO-	1	2	3	4	5	Etc
1	DOT	CIRCLE	SPHERE	Sphere4	Sphere5	Sphere6
2		DOT	CIRCLE	SPHERE	Sphere4	Sphere5
3			DOT	CIRCLE	SPHERE	Sphere4
4				DOT	CIRCLE	SPHERE
5					DOT	CIRCLE
Etc						DOT

Table (6.1): The painting of a 'dot' in various spatial 'DO'-dimensional spaces results in various instantaneous shapes in the respective higher 'SEE'-dimensional spaces.

Table (6.1) shows the effect of 'painting a dot' in various 'DO'-spaces of increasingly higher number of dimensions. The first data row is associated with a 1-dimensional 'DO' space (in which the dot is painted). The columns of this row then represent the various appearances of this dot in 'SEE'-spaces at increasingly higher dimensions. In a 2-dimensional space this dot will appear as a circle, in a 3-dimensional space as a sphere. It cannot be envisioned what the associated shape would look like in a 4-dimensional space (or higher). Therefore, within the table this shape has been named 'Sphere4'. And likewise there would be a 'Sphere5' and 'Sphere6', and so on.

The second row represents the appearances in the case where a dot is painted in a 2-dimensional space. One would than see a circle from the perspective of a 3-dimesional space, and a sphere from the perspective of a 4-dimensional space. And one would see a 'Sphere4' from the perspective of a 5-dimensional space, etcetera.

In all scenarios where just one single dot is painted in some n-dimensional space, the single fact that higher spatial dimensions exist has a consequence *within that same n-dimensional space,* the space in which the dot is painted:

> One other point is thereby instantaneously *virtually* pinpointed: the *Virtual Spot.*

This virtual spot also meets the requirements of the coordinates. But this fact *cannot* be noted from within that space itself, but only from the perspective of higher dimensional spaces. Its location can however be pinpointed: it is located at the mirrored position of the painted point (mirrored relative to all points of reference of the 'DO'-space). Communication between a selected location and it's uniquely associated *Virtual Spot* is instantaneous, based on the fact that painting a single dot is an instantaneous event.

The above can indeed –partially- be envisioned through the human brain, because the *spatial* appearance of the Crenel has been used. And –for human envisioning-space is limited to 3 spatial dimensions. The model in Crenel Physics is extrapolated beyond those 3 dimensions. From a Crenel Physics viewpoint (and also from a Metric Physics viewpoint), as was discussed before, there is no reason known why nature should restrict itself to whatever limitations may exist to the human envisioning of the appearance 'space'.

> The human *imagination* goes beyond the human *visualization* capabilities. One can *imagine* a 10-dimensional space and apply physical rules within it, without being able to *envision* this space. As one can *imagine* and use the unit 's^2' without being able to *envision* a square second.

And as discussed before, the above findings are not just applicable to the *spatial* or *distance* appearance of the Crenel: they –by implication- must be applicable to the underlying Crenel as well. Thus –by implication- the above finding must also be applicable to the *time* appearance of the Crenel.

The mechanism of 'instantaneous communication between any selected point and the associated and unique *Virtual Spot*' is applicable to any Fixture.

Through experiment, the higher than light speed communication has indeed been verified in Metric Physics. Crenel Physics embeds a potential underlying mechanism.

7. Cylinder model, physical constants and force.

One cannot visualize *motion* without visualizing at least one instance of a 1-dimensional *spatial* appearance and one instance of a *time* appearance of the Crenel. Motion can then be envisioned along a 'line' as 'time' is elapsing.

However, as was discussed, a 'line' does not necessarily represent a 1-dimensional spatial appearance. From a spatial 3-dimensional perspective for example, for a 1-dimensional space one should envision concentric spheres instead. But motion is not envisioned along such concentric spheres, but indeed along a 1-dimensional *line* embedded in this 3-dimensional space. The requirements for *visualization* of a process -like motion- thus seem tighter relative to just the visualization of a 1-dimensional space embedded in a 3-dimensional space. Such tightening of requirements is possible through *modeling*.

Through modeling one may also visualize Photon energy levels that are associated with the appearance of colored light (see chapter 13). One does not *see* the Photon energy level itself: instead one sees 'light' that appears to have a 'color'. It then takes a *model* to put this visible light into the perspective of a specific Photon energy on the Electromagnetic Spectrum, and to translate this to color information. Thus the model pinpoints the real physical parameter underneath and relates it to the subjective –human- observations.

*It is not **nature** that is modeled here. Instead, it is the subjective human perception that is modeled.*

Therefore, models do not necessarily represent the objective facts: they often are no more but *an effort* to explain observations within a human context. A model could be incomplete, have a limited focus or range, be an estimate, or even be fundamentally wrong. And nevertheless a model can be useful. The model of a flat Earth with the Sun orbiting around it was physically wrong, but nevertheless it was useful to humans.

Through modeling one can also *imagine* -rather than *visualize*- processes, for example by the usage of mathematical representations. As was discussed before, these mathematical imaginations can be seen as *virtual appearances*. Mathematics has the advantage that one can enhance, extrapolate *and* successfully apply rules beyond visualization: thanks to mathematics the human *imagination* capabilities go beyond the *visualization* capabilities. And models are the vehicle.

From a mathematical point of view one can for example imagine a 10-dimensional space, despite the fact that such a space cannot be visualized. And within a 10-dimensional space e.g. the formula to calculate a 'distance' between two 'points' can be derived by extrapolation from the well known (and visualized) procedures that are valid in a 1-dimensional space, a 2-dimensional space and a 3-dimensional space. Mathematics is full of such extrapolations. Here is the equation of a 'distance' between a point 'A' and a point 'B' in an 'n-dimensional' space:

$$Distance = \sqrt{\sum_{n=1}^{n=n} (X_A - X_B)^2}$$

It can be visualized and mathematically proven that this equation for 'distance' is valid for a 1-dimensional space, a 2-dimensional space and a 3-dimensional space.

Strictly speaking, mathematicians now reversed the reasoning. They say: 'we call the outcome of the above equation the *distance* between point 'A' and point 'B' in an n-dimensional space'. This is different from stating: '*distance* exists in any n-dimensional space, and is calculated according to this equation'.

The difference in formulation is paramount: from a mathematical viewpoint and according to the first definition, the concept 'distance' is nothing more but a *model*, accurately described by the given equation. There is no argument against defining 'distance' as a mathematical model.

Things become debatable however, when physicists associate a *physical property* to this mathematical model, and thereby de facto apply the second statement. The *physical property* 'distance' is a parameter for calculating e.g. gravitational force, magnetic force and electrostatic force. By substituting the mathematical model for 'distance' (given by the above equation) where in fact the *physical property* is required, they extrapolate. Such extrapolation is risky (but also tempting). It is like applying the (mathematical) Pythagoras theorem to a triangle in a plane, while one cannot be sure that this plane is 'flat'. In Metric Physics units of measurement, *no* plane is flat when mass is around. The above equation for distance is only valid in an empty (Euclidian) space. And therefore, in Metric Physics –dealing with mass/energy containing entities- it can only deliver *approximate* results. Crenel Physics must attempt to avoid such ambiguities.

In this chapter 'space', 'physical constants' and 'force' are further explored.

As discussed in the previous chapter, spatial coordinates can be split up into two types:

- The type of coordinates that express a *relative direction* (a relative angle like φ or θ), and
- The type of coordinates that express a *relative distance/time* (like 'r').

Figure (6.1) showed both types being used in spherical coordinates. These are shown up to a limit that one can still spatially visualize: a 3-dimensional space. They both are *relative* to some -encompassing- frame of reference. Within the shown frame, there is *one* point of reference (the point where the three axis cross), and there are *three* reference directions (the directions of the three axis themselves). From a modeling viewpoint however, there might as well have been 25 points of reference and 37 reference directions. The fact that such extrapolation cannot be visualized in a limited 3-dimensional space is not relevant, as long as through modeling some logical – mathematical- rules can be applied to the extrapolation.

Furthermore, in space there is no such thing as one *universal* reference point, one *universal* reference direction, or one *universal* reference time. All the associated coordinates therefore are relative to some Fixture which by itself is not fixed, and thus is 'floating around'. A Fixture just fixes the included coordinates *relative to itself*.

In practice, the reference points of spatial Fixtures need to be -directly or indirectly- associated with tangible objects. And should -for some reason- the coordinates of an object *within* that Fixture change, this always is a *relative* change within the Fixture. One may conclude that due to some coordinate change during some elapsed time, the object has moved. One may however as well argue that instead the Fixture moved. Because the coordinate change is a relative change, there is no objective way to positively tell the difference.

Low velocity, the 'cylinder model'.

It is from this setting that the only 'Package containing object' so far within Crenel Physics -the Photon- will be further investigated. The unique feature of this object is, that it cannot be 'grabbed'. Or in other words: it cannot be contained at fixed coordinates within *any* Fixture for some period of local time (whatever 'local time' may be…). This is, because Photons –by definition- refuse to give up their speed relative to any observer: this speed is –trivially- equal to the speed of Photons, under all circumstances and within any Fixture or frame of reference where time is propagating. Because light consists of Photons, the speed of Photons is also referred to as the 'speed of light'. The fact that Photons cannot give up their speed relative to any observer causes, that Photons do not fit into the human perception of the world. Here, they can only exist as a 'model' or as a 'concept': humans are capable of dealing with this actual Photon property (and the shear fact that Photons do exist) by applying *mathematical rules* that come forth from the property. Mathematical rules –by definition- are abstract rules, and Photons are abstract entities. Each time a human being envisions a Photon, he has to realize that it is not the real truth he is envisioning: he is envisioning an abstract model of a Photon.

Despite that Photons cannot be 'grabbed', one can imagine that it is nevertheless possible to spatially trap and contain a Photon in a confined space, e.g. in an imaginary 1-dimensional cylinder with a mirroring surface on the inside.

Within this containment the Photon still could not be 'grabbed'. Instead, it would bounce back and forth at its constant speed of Photons (relative to the container). And thus, relative to this container, the frequency of this bouncing does not depend on movement of the container itself. It is strictly depending on the spatial layout of such a container.

It can now be envisioned that this container initially is not moving relative to some encompassing Fixture that is set up by an external observer. This external observer will than notice the same bouncing process –from some remote distance and therefore slightly retarded- as an observer within the container would notice. Both would monitor the process of the Photon bouncing, and both would count the frequency thereof being at equal pace, presuming that the Fixture does not contain any other objects that could impact time measurements.

Note that there could be numerous remote observers, each presumably having their private Fixture in which the container is initially not moving. And –in empty space- all would agree on the equal bouncing frequency of the Photon within its container.

The assumption, that *if* in all these other encompassing Fixtures the container is not moving (= changing its coordinates), this does *not* imply that these encompassing Fixtures are frozen relative to each other. A Fixture could rotate around the cylindrical container (according to others), while the container itself is the axis of rotation. Within this apparently rotating Fixture (apparent to others) the coordinates of the container would not change, while the encompassing Fixtures nevertheless observe a relative rotation between them.

Things become different, if one remote observer -just temporarily- *accelerates* the container within his spatial Fixture. This *must* cost force because –although the imaginary container itself is presumed weightless- the Photon therein contains Packages. Therefore, equation (2.1) is applicable: F=m.a, in which variable 'm' represents the *mass appearance* of the Photon's Packages. While the force is applied for some limited period of time, the thus accelerated container -and the Photon therein- *must* absorb an amount of energy delivered by the external observer. Based on dimensional analyses, the amount of delivered energy is equal to the strength of the force, multiplied by the spatial length of the path along which it was applied.

In Crenel Physics this absorbing is facilitated trough an increased Package containment of the Photon. This would thereby result in a numerical increase of *all* appearances that are associated with Packages: frequency, mass, energy containment, strength of gravity…

Here this energy transfer is re-investigated using the current cylinder model: how can this energy transfer be envisioned?

The presumed universal constancy of the 'speed of Photons (or: light)' plays the key role in the answer.

> It is at the bottom line not relevant whether the 'speed of light' is objectively verifiable for being a universally constant, or not. It is only relevant that this value is used as *reference* for a series of related units of measurement. The statement that 'the speed of light is a universal constant' may snow under what thereby exactly is stated here. To acknowledge the bottom line, the statement

can be reformulated in *two* underlying rules:

1. 'Photons always travel at the speed of Photons'.

One cannot argue with this –trivial- rule.

The essence is that this Photon speed –whatever it may be- will be used as a *natural and universal standard*, usable in any frame of reference. This usage is also possible due to the second rule:

2. Photons do not ever overtake each other (while traveling through equal space).

Under given circumstances all Photons travel at *equal* velocity. It is this latter fact that indeed makes the Photon velocity special, and usable as a standard. It is however impossible to prove that this statement is true anywhere and anytime: there are infinite numbers of types of spaces and local conditions thinkable. However, as far as known, this second statement has hold in all cases so far.

Can –given the above- the *aftereffect* of the temporary applied force (and thereby the temporary acceleration) be noticed by the observer who is residing *within* the container?

This internal observer will still see the *same* original container. And within it the Photon is still bouncing at the velocity of Photons. It is not relevant if there could be some hypothetical *objective* means to identify a potential velocity change of the Photon: the velocity of the Photon

–whatever that would be- *is* selected as reference. It is for this reason, that the locally observed frequency of the Photon remains unchanged *by definition*. Also, it thereby did not matter in which direction the container was accelerated. The internal observer will just notice the encompassing Fixture -in which the container stood still originally- now moves at some velocity.

This is a remarkable conclusion because it can be generalized: *any* transfer of energy to a Photon in a container will -within its local world- *not* be noticed. From the local perspective one would only notice a change in the encompassing Fixture. Once a Photon is born, its *internal* world is insensitive to externally applied forces. It is the encompassing Fixture that may transfer some energy to or from a Photon, and it is the encompassing Fixture that will see the consequences. The Photons local world behaves -so to speak- autistic. It can be pushed around without internal consequences.

The question then is: how does the encompassing Fixture (or: the encompassing world) notice the impact of the temporarily applied force?

Relative to the *remote* observer, the Photon within the container also -by definition- remains moving at Photon velocity. It is based on the second statement (Photons cannot overtake each other) that it is not relevant that the container itself is now moving relative to the encompassing Fixture. Note however, that at closer look the respective *times* required for the to- and fro-paths (associated with the bouncing) will –from the remote perspective- shift. The total round trip time –associated with frequency- is not impacted by this shift. In Crenel

Physics this effect will be named *longitudinal polarization*: from a remote perspective the wave geometry of the Photon's bouncing pattern is not symmetrical anymore in the direction of the movement.

> *Longitudinal Polarization* is:
> the a-symmetric wave function in the direction of movement.

By elimination, for the external observer there is only *one* parameter left to reflect the energy absorption: according to the remote observer, the *spatial size* of the container now must appear *shrunken* relative to its original size. This container size is the only remaining parameter in the model.

> In this experiment there are three potential parameters:
>
> - The energy that is transferred to the container (to accelerate it),
> - The spatial size of the container, and
> - The velocity of the photon contained within it.

The experiment requires a relationship between these three parameters based on the principle that 'energy cannot be lost'. Because -by choice- the third parameter (the velocity of the Photon) is set fixed, the spatial size of the box must fully reflect the energy transfer.

Note thereby, that there would have been no

fundamental objections against -alternatively- defining some fixed 'universal box size', and use this fixed size as the standard for the other units of measurement. This alternative approach would than require that –in this case- a change in the Photon velocity would reflect the energy adsorption to the remote observer. This would – obviously- lead to an entirely different basis for *all* units of measurement. Nature would then reveal its rules through other relationships.

In conclusion: Crenel Physics *by choic*e follows the approach of Metric Physics, where the Photon velocity is set as the reference.

And consequently, with equal Photon velocity in a shrunken container, the *remote* observer must see the Photon bounce back and forth at a *higher* frequency than before. This makes the above model consistent with Planck's law ($E = h.\upsilon$). Here, higher Photon energy is proportional to a higher Photon frequency. And furthermore, the remote observer may notice the aforementioned *longitudinal polarization*: the a-symmetry in the to- and fro-times of the bouncing.

In the aforementioned Planck's law it is the Photon *itself* that is associated with a frequency. In the above model, at 'lower than light velocities' it is the apparent shrinking of a container *around* it, which dictates a frequency shift. This shrinking is *apparent*, because it would not be observed from within the container itself.

The above experiment described what happens if one traps a Photon in some imaginary container, and manipulates this container with an external force. The model allows the associated container to be *held in place*, or move around at *limited velocity*, without violating the rule that the Photon itself maintains its –universally constant- Photon velocity.

It is a logical step to now explore extrapolation of this model, and assume that even in absence of this imaginary low speed containment, each Photon is nevertheless and *always* trapped in a likewise *private* containment that is going where the Photon goes, and which thus is also traveling at Photon (or: light) speed. And within this private containment, the Photon itself is *still* moving around at Photon speed. The Package containment of the Photon is than to be associated with the size of this private box, rather than with the Photon –being no more than an abstract concept- residing within it.

There is evidence in nature, that the above model reflects the actual properties of a Photon. Through the force of gravity, energy can be transferred to and from a Photon through the eyes of a remote observer. E.g. a Photon leaving the Sun will lose some energy through the Sun's gravitational pull. As the velocity of the Photon itself is set constant, one can now -trough this model- envision that through the eyes of a remote observer the Photon's private container is widened through the energy loss, resulting in a lower frequency.

The originally used cylindrical container (that initially was not moving relative to a remote observer) could not possibly be accelerated all the way up to light speed. As the container accelerates, its size would shrink and thus the frequency (and thereby Package containment) of the contained Photon would increase proportionally. Therefore, assuming a constant force over time, the resulting acceleration of the container would drop as the container shrinks in size. At light speed, the container would appear infinitesimally small to the remote observer, and the Photon's Package containment would appear limitlessly high. The appearance *mass* would be infinitely large. Therefore, further acceleration would cost infinite force. The container could approach light speed relative to a Fixture, but never reach it.

One can also look at this speed limitation from another perspective: the aforementioned *longitudinal polarization* effect. As the container gains speed, this polarization effect (the a-symmetry of the bouncing function itself, observed by the remote observer) would increase, while the total round trip time itself would not be impacted. Because Photons have equal velocities in all directions, an *increase* in to-time (observed by a remote observer) would therefore be exactly balanced by a same amount of *decrease* in from-time. Because neither of the times could become negative, this inherently sets a limit to how far this longitudinal polarization could go, and thereby sets a maximum to the possible velocity of the container: full longitudinal polarization (where either the to-time or the fro-time has become a value of zero) is the extreme possible, and consequently dictates a limit to the maximum container speed.

Nevertheless, Photons themselves *do* have this 'impossible to reach' light speed. Physics apparently splits nature into two types of objects:

1. Objects that do have light speed (Photons). These objects can have *no other* speed than this light speed, and this speed is used as a universal standard to any observer.

2. Objects that do *not* have light speed. And these objects will *never* reach light speed relative to any observer.

The model of a Photon can now be further fine tuned as follows:

> *A Photon:*
> *is an object propagating at the speed of Photons relative to any frame of reference. It carries an imaginary private spatial container around it in which it is bouncing back and forth with the speed of Photons. The -relevant- spatial size of this imaginary container is equal to half of its wavelength.*

The amount of space that this '*private* Photon container' occupies is reciprocally related with the amount of energy that a Photon represents:

- The earlier defined 'zero-Photon' is associated with an infinitely large private container,
- The more Packages a Photon contains, the smaller its private containment.

This model gives the Photon a *spatial* appearance. And it is this spatial appearance that also represents the amount of energy containment. The model thus implicitly gives a *reciprocal* relationship between this *spatial* appearance and the *energy* appearance of a Photon. And thus, it gives a *reciprocal relationship* between the underlying fundamental physical properties: the Crenel property of the spatial box, and the Package property of its containment. This reciprocal relationship will be reconfirmed later, trough another approach.

This raises the question whether the fundamental physical properties that have been defined in Crenel Physics (being the Package and the Crenel) indeed are *independent* properties. It so seems, they are not...

Investigating Mathematical rules in Crenel Physics.

When using spatial appearances, it is relevant to have insight into the validity of mathematical rules.

The frame of reference represented by figure (6.1) presumes a 'linear' 3-dimensional space or 'Euclidean' 3-dimensional space, in which e.g. the circumference of a circle equals $2.\pi$ times the radius of that circle, and in which also e.g. Pythagoras' theorem for any rectangular triangle is valid. This also lies at the basis of the 'radial', being an absolute unit of measurement for angles.

In Crenel Physics, distances are measured in 'Crenels' along the path that Photons would take. The numerical outcome is based on the –universal- speed of Photons (or:

light): Photons *are* light, and *thus* have the speed of light. One thereby uses a clock to quantify the time required for Photons to cover a distance. This time is expressed in Crenels, and the outcome *is* the distance.

In Metric Physics, when 'Packages' are somewhere around in the area (and thereby there is gravity), time measurements are impacted and space appears curved. Therefore, from a remote perspective Photons would not exactly follow 'straight' paths relative to a metric grid ('straight' as defined by the mathematics presumed in figure 6.1). And consequently, space is not always 'Euclidean' anymore.

Or: when distances are expressed in meters and when mass is around, the circumference of a circle will *never* equal exactly $2.\pi$ times the radius of a circle. This is because the circle is *not* embedded in a 'flat' plane (or 'un-curved' space). And Pythagoras's theorem would not hold either. This cannot be visualized by the subjective spatial appearances that humans perceive. One can however apply models (mathematical rules) to retrieve the correct compensations for these cases.

However, when distances are expressed in 'Crenels', the path to follow between two points *is* the path a Photon would take. And one would define this path as being 'straight' by definition (or: in Crenel Physics this is *the* criterion for determining straightness). The grid within the frame of reference is based on these paths. Thus, the *experienced* space is by definition - from a *physical* viewpoint - 'straight' and 'Euclidean'. When expressed in 'Crenels' the circumference of a circle in this Fixture would *always* be 2.π times its radius. And Pythagoras' theorem will *always* hold.

To gain more insight in the difference between Metric Physics and Crenel Physics, imagine that a Photon is traveling a triangular path that is perpendicular to –and in the vicinity of- the surface of a very heavy object, e.g. the Sun (see Figure 7.1). The corners of the triangle would be the spatial points 'A',' B' and 'C'. Assume that at each corner point there is a beacon light for convenience, visible from all directions.

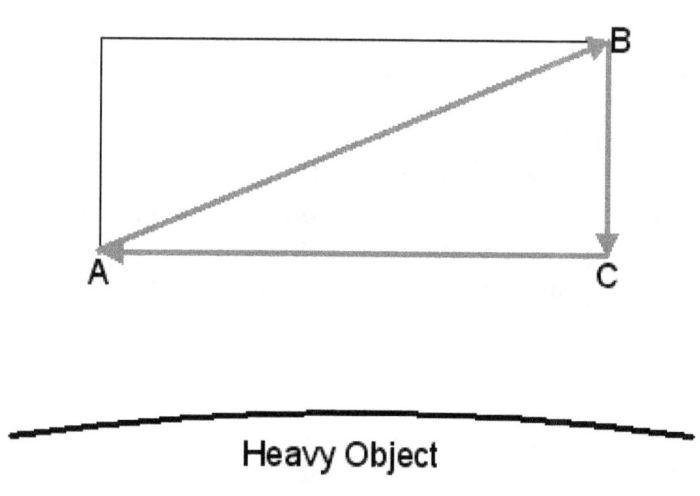

Figure 7.1: Triangular Photon path above surface of a heavy object

One experiment now would be to give the Photon traveling instructions, such that it will depart from point 'A', than travel via 'B' to 'C' and then back to 'A' again. And that the second objective is, that the round trip is to be done as fast as possible. As in both Metric Physics as well as Crenel Physics the Photons velocity is a maximum velocity that serves as a universal standard for units of measurement -and therefore it cannot be seen as a variable- this second requirement boils down to the instruction: touch points 'A', 'B', 'C' and 'A' along the *route a Photon would take*.

The experiment then is, to review the required travel instructions, from various viewpoints.

There is a 'remote' viewpoint (from any distance, being a 'large' distance relative to the size of the triangle) and a 'local' viewpoint (at the Photon itself). And there are viewpoints per Metric Physics, as well as per Crenel Physics. The following four perspectives for route descriptions are of particular interest:

1. Metric Physics instructions from a local perspective,
2. Metric Physics instructions from a remote perspective,
3. Crenel Physics instructions from a local perspective,
4. Crenel Physics instructions from a remote perspective,

Regardless these various perspectives, the *outcome* for the route description to deliver to the Photon must expectedly result in the same unique physical route: one must assume that there is only *one* unique solution…

From the remote perspectives, due to the gravitational field the Photon will lose energy while climbing the sloped section AB. Instead of losing energy through loss of velocity (as a stone would do), it will:

- According to Metric Physics lose frequency (or mass), and,
- According to Crenel Physics it will lose Packages (to be reflected in the various appearances of the Package).

In both cases however, the forward Photon velocity will remain –by definition- constant from any perspective. The lost energy on the way up will be fully recovered while coming down along the path BC.

Perspectives 1 and 2 (Metric Physics):

Here, the coordinates in figure (7.1) are given in metrical units of measurement, within a metrical spatial grid.

The first instruction to the Photon would be the *direction* to depart from at point 'A'. This should *not* be the direction straight towards the metric coordinates of point 'B'. This would lead to undershooting point 'B', as the Sun's gravitational field will accelerate the Photon downward, into a direction that is *rectangular* to its forward path (note that this is *not* a vertical direction of acceleration, because such direction would also slow down the Photon):

> *Photons –always maintaining their constant velocity- can only be accelerated into a direction that is perpendicular to their forward velocity (!).*

Therefore the instruction will need to be, to aim somewhat *above* the metrical coordinates of point 'B'. It has been discussed in previous chapters, that thereby the *initial* energy -or mass containment- of the Photon used, will *not* influence the path that the Photon will physically follow, as long as it initially contains enough energy to reach the top 'B' of the triangle. This path is equal for *all* Photons, heavy or light. Thus the first instruction, the angle of departure, would be equal for any Photon.

There is a practical procedure to avoid having to do a precise angle calculation: from the perspective of starting point 'A', the beacon light of point 'B' is not seen at the metrical coordinates where it actually *is*. This is because Photons that depart from the beacon light at point 'B' will undergo the exact same curving: they follow the same curved path, in opposite direction. Thus, their impact angle while arriving at point 'A' is exactly the angle to shoot into.

Therefore, in scenario 1 the instruction will be to depart into the direction towards the coordinates where point 'B' is *seen*, to its *apparent* position. Within the metric grid this will be a point that is located slightly above the coordinates where point 'B' physically is.

In scenario 2 (the remote position) this pragmatic shortcut is not an option, because the local visual information from point 'A' is not available. Thus the remote observer will recognize that the Photon will follow some curved path through the metric grid, and will have to reconstruct this path through calculation.

Thereafter it is just a matter waiting, because the departed Photon will –after a certain amount of elapsed time- arrive exactly at the physical position of point 'B'.

The second instruction to the departing Photon needs to be, *how much time* -expressed in seconds- it exactly should take until arrival at 'B', where a course change is to be made to 'C'.

Here, the local observer at point 'A' again has an advantage: he could send a test Photon to 'B' and have it returned back by a mirror at point 'B'. Thus he would conclude that the travel time will be exactly half the time of the round trip: to and fro, this test Photon will follow exactly the same path, and all the way its forward velocity remains constant. Therefore, the second instruction to the Photon would be, to follow the initially given direction for exactly half the round trip time.

However, one complication thereby is that this timeframe would be based on a clock which is positioned at point 'A'. The clock at 'B' runs at faster pace, as it is located further away from the Sun's central point of gravity. And the clock of the observer at some remote location 'X' runs even faster.

This is because in Metric Physics, time measurement is found to be relative to local gravity.

Thus, where 'A' would locally measure a round trip time of e.g. exactly 1.00000000 seconds, from 'B's perspective the clock at 'A' is running at slower pace. When 'B' would measure the exact same round trip time, according to his clock it would last longer and result in a measured timeframe of e.g. 1.00000018 seconds.

At the remote location 'X' the 1.00000000 seconds measured by the clock at 'A' would correspond to an even longer local timeframe of e.g. 1.00000050 seconds. Likewise, the measured timeframe of 1.00000018 seconds measured by 'B' also would exactly correspond to this same 1.00000050 seconds at 'X'.

Because there must be one unambiguous second instruction for the Photon to follow, according to Metric Physics nature must house a mechanism that reflects this predictable and consistent difference in clock paces. Some rule must exist that unambiguously defines one unique travel time instruction to deliver to the Photon upon departure from 'A' towards point 'B'.

Given the scenario, and recognizing that the difference in time pace is indeed pending local gravity (it can nowadays be confirmed through experiment, using very accurate clocks) while the Photons velocity is presumed equal and constant from all perspectives, this would imply that the 'meter' -as used in the Photons velocity indication- (velocity is expressed in meter/second) must be *exactly* as pending on local gravity, as the 'second' is. In other words: because –relative to some remote location- the 'second of time' measured within a gravity field appears to last *longer* (the remote clock in the gravity field appears to run slower), so appears the meter *stretched*. Only through this exact and equal dependency on gravity, the ratio in which velocity is expressed (the ratio meter/second) remains independent from gravity.

This equal and proportional dependency of both 'time' and 'distance' on gravity is proof that they *cannot* be seen as non-related units of measurement. Note, that in Crenel Physics they both *are* quantified using the same unit of measurement: the Crenel.

Over all it is clear that the challenge to give precise travel instructions to the Photon is quite complex using Metric Physics coordinates and units of measurement.

Perspectives 3 and 4 (Crenel Physics):

Here, the key difference is that the spatial coordinate grid associated with the spatial frame of reference -in which triangle ABC resides- is using Crenels as yardstick. And thus 'straight lines' are based on the path that Photons would take. When these paths are drawn as gridlines, these lines would also *mathematically* appear straight. Mathematical rules like Pythagoras's theorem are valid within it. It is a 'natural' grid.

Using the coordinates of this grid, from the perspective of point 'A' the physical location of point 'B' (or: its Crenel coordinates) matches the location where it is visually *seen*. This simplifies the procedure to give directional travel instructions to the Photon that is about to start its round trip along the triangle (at what angle to depart, what angle to turn upon arrival at point B towards C, etcetera). One would not need a beacon light.

The second instruction –again- involves the *time* that the Photon would need to travel from 'A' to 'B'.

In *none* of the four scenario's one would need to give a clock to the Photon to allow it to keep track of covered distances. Instead, one can use the aforementioned 'private container' that is associated with the Photon, and in which it is bouncing back and forth. Thus, instead of using a clock, it would be more direct to instruct the Photon to keep moving in the specified direction for a certain specified *number of bounces* (and possibly fractions thereof). This would be a *dimensionless* number. By using this feature, the tracking of 'time' has become obsolete… and is replaced by a simple counter.

Because this number is dimensionless, it does not matter if –for a given Photon- this number is calculated through Metric Physics or through Crenel Physics: the numerical outcome must be the same.

Obviously, the number to specify will be proportional to the photon's initial container size at point 'A': a high energy Photon would reside in a smaller container, and thus receive a higher number of bounces to travel in the specified direction, relative to a low energy Photon. This is, because both the high energy Photon and the low energy Photon travel at exactly the same velocity.

Between Metric Physics and Crenel Physics there is however a difference in evaluating the container size: it is expressed in 'meters' or in 'Crenels' respectively.

> In Metric Physics and from a remote perspective at location 'X', along the path from 'A' to 'B' the Photon keeps its original velocity (in meters per second) whereas the clock attached to the Photon appears to run at faster and faster pace as the

Photon climbs in the gravity field. Thus, 'X' would reckon that the local meter –where the Photon is- must be shorter relative to a meter measured at the remote position 'X', based on the fact that the local velocity (expressed in meters per second) still is the same constant. At the same time the remote observer 'X' will notice the bouncing frequency drop according to his clock. This remotely observed dropping of frequency suggests that the Photon is now bouncing in a larger sized box (expressed in remote meters), whereas at the same time the applied meter seems shrunken. Thus, the remote observer will conclude that both relative effects combined compensate each other when translated to the local world of the Photon. The remotely observed time contraction is compensated by a remotely observed proportional space contraction: the ratio 'meters per second' (in which velocities are expressed in Metric Physics) is at the bottom line a constant. This is consistent with the earlier finding, that external forces have no impact at all when evaluated from within the Photon's container: along the path from 'A' towards 'B' its size appears to remain constant, and the Photons bouncing remains unchanged. Even when distances are expressed in meters, and time is expressed in seconds. In Metric Physics it therefore is the challenge to determine the length of the curved Photon path between 'A' and 'B': curved through the metrical grid. That leads to quite complex calculations that quickly would gain complexity if the gravitational field itself has a complex shape. This would for example be the case if a Photon would travel through a complex molecule.

The imagination of velocity in Metric Physics would be simplified, if each velocity was measured as a fraction of the light velocity. For example: an object 'Q' moves at 0.000123 times the velocity of light. Because the velocity of light is presumed constant within any frame of reference, one could safely argue that this fractional velocity of 0.000123 times the velocity of light is also the same velocity from the perspective of any frame of reference. The fact that meters are perceived totally different from seconds, allowed humans to create a *perceived* reality based on these different perceptions. This form of subjectivity is compatible to the human ability to perceive colors based on color perceptions: physically these do *not* exist (see chapter 13). What then remains after this subjective envisioning is the complexity in determining the length of paths to follow.

In Crenel Physics the two parameters 'time' and 'distance' are based on one single underlying unit of measurement: the Crenel. Thus, velocity is expressed in Crenel per Crenel, and this ratio is *obviously* immune for any potential impacts to the Crenel. Even if the Crenel itself would appear to be a relativistic unit of measurement, this would have no impact on the numerical value of a velocity itself: this number remains what it happens to be. Because the spatial grid is also expressed in Crenel, a constant velocity –like a

Photon velocity- can be graphically drawn in as a vector with constant length. Because the path between 'A' and 'B' is expressed in Crenel, through a simple division of lengths the required traveling time can be found (expressed in Crenel). And in turn this can be translated into a number of bounces that would take place within the Photons container which has a fixed size from its local perspective. The thus found number can be given to the Photon as the aforementioned second traveling instruction.

And because time measurements take place in Crenels, there are no pace differences between clocks at various locations under the influence of gravity. Gravity is thus indeed no more than just one of the *appearances* of the Package: the appearance was named 'Strength of Gravity'. In Crenel Physics this appearance results in an exclusive force between any pair of objects, and in nothing else. It has no impact on other parameters, or units of measurement.

Although the advantages of Crenel Physics may seem only of conceptual relevance at first sight, they can be of paramount importance and help while describing fundamental physics. As the selection of a spatial coordinate system (rectangular, spherical, cylindrical, etc.) will have major impact on the complexity of modeling, so has the portfolio of the used units of measurement.

From Crenel Physics, a different definition of the concept 'velocity' comes forth:

Definition 4: 'Velocity'
Velocity = a dimensionless number that expresses the rate of Crenel conversion from one appearance (the 'time' appearance, 'Crenel$_T$') into another appearance (the 'distance' appearance, 'Crenel$_D$').

This definition makes understandable (perhaps even trivial), that such a conversion rate (or velocity) cannot be less than 0, nor more than 1:

- At a value of 0, there simply would be no conversion.
- And, a numerical value of 1 implies *full* conversion of one appearance (the 'time' appearance) into another appearance (the 'distance' appearance). It seems logical that there cannot be more conversion than *full* conversion. So that sets the maximum limit to 1.

Note: in Metric Physics the maximum possible speed is the speed of light in vacuum (approximately 300,000,000 m/s relative to the observer) which limitation seems rather 'unexplained'. This maximum is presented as 'one of those facts in physics that one should not try to understand'. In 'Crenel Physics' however, the maximum speed limit of 1 appears in a more inherent and intuitive manner. And because space becomes Euclidian, mathematical rules present exact physical realities: the circumference of a circular orbit *does* equal π times its diameter.

To further explore physical constants, the formulas (2.3) and (2.4) from Metric Physics (see chapter 2) can be combined into one single equation as follows:

$$E = m.c^2 = h.\upsilon \qquad\qquad (7.1)$$

Where:
E = energy in Joules (J)
m= mass in kilograms (kg)
c = velocity of light in meters per second (= 2.9979 x 10^8 m/s)
h = Planck constant in Joules x second (= 6.6256 x 10^{-34} J.s)
υ = frequency in number (of rotations) per second (s^{-1})

The first chapter in this book claimed that Metric Physics uses more fundamental units of measurement than the number of fundamental physical properties would justify. This can also be expressed in a mathematical way as follows: the equal sign '=' which appears twice in equation (7.1) seems not to be taken very literary.

The equal sign '=' mathematically connects two sides of an equation, and expresses that the left side of the equal sign is *exactly* equal to the right side. The left side is *not approximately* the same as the right side. Also, it is *not* to be seen as an alternative to the right side. Both sides *are* equal and the same. Formula (7.1) contains *two* equal signs, which connect three parts to each other. Therefore, all three parts are to be seen as being *exactly* the same, without any compromise.

According to the '=', not only the *numerical value* of each side of the equation is equal. The *units of measurement* (or combination thereof) of each side are equal as well: one cannot set apples equal to pears.

Although three equal parts are connected, the units of measurement in which the three parts of equation (7.1) are expressed house no less than *four* different fundamental metrical units: the Joule, the kilogram, the meter and the second. And furthermore the equation houses *two* universal physical constants: the velocity of light in vacuum 'c', and Planck's constant 'h'. The role of these universal physical constants in equation (7.1) is, to act as a *conversion factor*.

By evaluating equation (7.1) and by considering the fundamental physics behind it, it was concluded in chapter 1 that only *one* physical unit of measurement is required here: the 'Package'. In Crenel Physics equation (7.1) then translates to the following:

$$\text{'Package}_E\text{'} = \text{'Package}_M\text{'} = \text{'Package}_F\text{'} \qquad (7.2)$$

Equation (7.2) is not trivial although this may appear so at first sight: it de facto expresses that the 'Package', this fundamental physical property, can profile itself through three different *appearances*. In this case an *energy* appearance, a *mass* appearance and a *frequency* appearance respectively. In equation (7.2) these different appearances are indicated by the subscripts 'E', 'M' and 'F'. According to equation (7.2) these different appearances all three *are* the same.

Furthermore, an additional appearance of the Package was introduced: the 'Package$_{SOG}$': it expresses that when Packages are around, a proportional 'Strength of Gravity'

123

is also around. Albeit that a gravity force only and exclusively appears between *pairs* of objects. Only by use of an object that contains Packages, the 'Strength of Gravity' of another object –subject to the investigation- can be found.

The human brain is not trained to accept the equality of the various different appearances: how can these represent 'exactly the same' while they appear so different?

Humans base their thinking -and thereby most sciences- on *appearances* for practical reasons. It is the appearances that are perceived. And therefore humans have to primarily deal with these, and not so much with their bottom line.

As already indicated, chapter 13 describes 'colors' as being one of the *appearance* outcomes of Photons. It illustrates that an analytical approach with regards to appearances is relevant, but also that 'colors' are dictated by -and perhaps limited to- the subjective human perception. While humans can enjoy art that is based on colors, it should be recognized that this is a privilege which is not universally shared. Not even amongst all human individuals (some are blind).

By going back to the basics, humans learned to produce and reproduce the appearance of colors. Colors that in fact do *not* exist at the bottom line of physics: light particles (Photons) are *not* colored. The physics behind living in a colorful world is only 1-dimensional: the electromagnetic energy spectrum of Photons.

Likewise, *sound* is an important but subjective appearance to humans. Humans can enjoy an appearance like e.g. the sound of music, while this same sound is at the bottom line nothing but an accumulation of noises produced by instruments, which in turn produce a series of accumulated pressure waves that hit our ears.

Will -by focusing on the bottom line in 'Crenel Physics'- the music be taken away from science? Will one be stuck with pressure waves only? Certainly *not*: thanks to these fundamental insights, humans learned to create and manipulate appearances where this must have seemed impossible before. It is hard to appreciate and remember the inventions that could only come *because* fundamentals were discovered.

> Just imagine that one would need to explain to a medieval knight that the song from a minstrel is actually shaking air that can be reproduced by a vibrating piece of cardboard (as he would describe a loudspeaker): one cannot explain this, unless the physical fundamentals of sound are addressed... This knowledge is essential e.g. for recording music, and to reproduce it through multi channel sound systems at high fidelity, whenever wanted.

As a human being, one does not need to (nor want to) adapt its brains to the fundamental physical facts alone, thus taking away the music and the colorful world. On the contrary: humans want to use knowledge about 'fundamental physical facts' to serve practical benefits. The fundamental knowledge helps humans to create appearances that make feel good. And that is –besides existential questions- another justification for Crenel Physics.

> Whether one creates a camp fire, a high tech multimedia device, or a painting: it is the *appearances* that reach humans, *not* the fundamental physical properties. *Humans live in an apparent world.* Humans *want* to live in an apparent world.

As discussed earlier, 'time' and 'distance' likewise are subjective *appearances* to humans. No more, and no less. Like 'colors', they do not exist as such. They are *appearances* of the Crenel. Horses may perceive entirely different appearances of the same Crenel.

A question is whether human beings - when they are born - are immediately aware of the aforementioned appearances like time, distance, music or color. Or does one gradually perceive these while growing up? Are these appearances the consequence of the design of human sense organs that developed together with the individual human brain, or are they 'for real'?

In 'Crenel Physics' the answer is, that the appearances that humans perceive since birth are *not* for real: they are the consequence of the design of their sense organs, in combination with subjective interpretation through the human brain. And these appearances become more apparent as the individual human body and mind develops. One consequently must accept that other living creatures –such as horses- may live in another *apparent* world. But at the same time: they *do* share the underlying *fundamental physical properties* with humans, even if they may not be conscious of this.

Persons who are deaf or persons who are (color-) blind live in another apparent world, which they nevertheless share with the non deaf and with the non blind.

Crenel Physics deals with the 'fundamental' world underneath:

> *The 'fundamental' world is the world stripped of its appearances. It is the world that remains when human sense organs – and associated appearances - are not taken into account, and all memories of these are ignored.*

This fundamental world is based on universally shared values. These will be further reviewed from here onwards.

The velocity of light 'c'.

Equation (7.2) seemed a trivial equation in Crenel Physics because it uses only *one* unit of measurement. The depth behind it is the recognition -coming straight from Metric Physics- that there are (at least) three different appearances of the 'Package' to humans. In Crenel Physics, a fourth appearance was added: 'Strength of Gravity'.

In equation (7.2) one does *not* need the universal physical constants 'c' -in its role as conversion factor- anymore, as in equation (7.1). This is because in Crenel Physics the unit of measurement for both the 'Energy' appearance as well as the 'Mass' appearance is the same: the Package. Therefore, between the first and the second part of the equation there *is* no requirement for unit (of measurement) conversion, nor for numerical correction. This can be rephrased as follows: in equation (7.2) the conversion factor for 'c' (in equation 7.1 found as 'c^2') has become a dimensionless number with the numerical value of 1.

> *In Crenel Physics, Photons travel at the speed of 1.*

The symbol 'c' will also be used for this constant:

'c' = 1

Planck's constant.

This constant bridges the first and second part of equation (7.2) to the third term. As the velocity of light ('c') has been seen as a conversion factor between the first and the second term of the equation, likewise Planck's constant ('h') is also to be seen as a conversion factor between appearances.

To further investigate the meaning of this conversion factor, the earlier definition for a 'Package' needs to be reviewed:

> *Definition 1: 'Package'*
> *Package = the physical property of a 'thing', expressing its quantity.*

Based on equation (7.1), in which Planck's constant 'h' represents the conversion factor between frequency (in metric units) and energy (in metric units), there now seems an argument to quantify one Package in Crenel Physics as follows:

> *Definition 5: 'One Package'*
> *One Package represents a quantity which corresponds to a frequency of exactly 1(one) full cycle per 'Crenel'.*

This definition has consequences because it consumes a degree of freedom in Crenel Physics: in the above definition the 'Package' (as a unit of measurement) is now *related* to the other unit of measurement: the 'Crenel'. Where – until here – the 'Package' and the 'Crenel' were presented as being two *separate* and completely independent fundamental units of measurement, definition 5 takes away this independency.

Through the above definition the true physics behind Planck's universal physical constant 'h' becomes apparent, and it is promoted to a fully integrated part in Crenel Physics. Planck's constant will – as a consequence to definition 5 - be a fundamental universal constant with the *numerical* value of 1. Had one – arbitrarily- decided that 'one Package' would correspond to a frequency of 12.34567 cycles per 'Crenel', the numerical value of 'Plank's constant' would also have been 12.34567. Up until here, while defining Crenel Physics, one still has the choice, the degree of freedom to settle Planck's constant… But once settled as done: that's it! Definition (5) settles and fixes the relationship between 'Packages' and 'Crenels'. It seems logical to settle for a numerical value of 1. This choice will save battery power in future calculations.

Definition (5) in Crenel Physics *is* Planck's law (written as E=h.υ in the Metric Physics).

This full integration – and anchoring - into Crenel Physics is a good idea, because Planck's constant 'h' indeed expresses a fundamental physical fact: the fact that 'Energy' can appear as 'Frequency'. The explanation why in Metric Physics the required conversion factor 'h' has such a strange unit of measurement (it is expressed in J.s) and such a strange numerical value (6.6256×10^{-34}) is: the metric units were already defined and calibrated *before* Planck could share his insight with the world. Fortunately, in Crenel Physics one does not have this historical problem.

The units in which Planck's constant is expressed in Metric Physics (J.s) can be easily translated into the units of Crenel Physics:

- the J(oule) relates to energy, and therefore is represented by the unit 'Packages',
- the s(econd) relates to time, and therefore is represented by the unit 'Crenel'.

Thus, in Crenel Physics, Planck's constant has numerical value 1, and is expressed in 'Packages' x 'Crenel'.

The following example illustrates the enormous impact of definition (5):

> If one takes (in Crenel Physics) a 'thing' with a quantity of e.g. 123.4 Packages (which is the unit of measurement for mass/energy), this same 'thing' also appears -amongst others- also as 123.4 cycles per 'Crenel'. Because 'cycles' is a dimensionless number, the unit of measurement of the right term of equation (7.2) is 'Crenel^{-1}'.

In other words: the unit of measurement 'Crenel' is the *reciprocal* of the unit of measurement 'Package' (!).

> *Where earlier the 'Package' and the 'Crenel' were called the Yin and the Yang of Crenel Physics, they are **not** to be seen as opposites or counter poles to each other… as the old Yin and Yang wisdom says. They are **reciprocal** to each other.*

> *Definition 6: 'Package versus Crenel'.*
> *The unit of measurement 'Package' is the reciprocal of the unit of measurement 'Crenel'. And vice versa the unit of measurement 'Crenel' is the reciprocal of the unit of measurement 'Package'.*

This is a fundamental result in 'Crenel Physics'. As will be shown later this also results in other interesting relationships. For example: the 'Crenel Physics' translation of Einstein's famous equation $E=mc^2$ (in

Metric Physics) can actually be based on Planck's law (or vice versa).

Or in other words, it will appear that:

- Starting the reasoning from 'Crenel Physics': because of definition (6), the outcome in Crenel Physics boils down to the equation $E=mc^2$.

- Or the reverse: starting to reason from 'Metric Physics', because here $E=mc^2$, consequently in 'Crenel Physics' the 'Crenel' is the reciprocal of the 'Package'.

  Or: 1'Package' x 1'Crenel' is equal to the dimensionless number 1 (one). This result is - remarkably- equal to the speed of light, which *also* has the dimensionless value of 1.

Here is a summary of units of measurements and related findings so far in 'Crenel Physics':

1. 'Packages' express the quantity of a 'thing'. From here onwards the symbol 'P' will be used for the unit 'Packages'. 'Packages' have various *appearances* to humans: 'Package$_E$', 'Package$_M$', 'Package$_F$' and 'Package$_{SOG}$', where the subscripts stand for 'Energy', 'Mass', 'Frequency' and 'Strength of Gravity' respectively.

2. 'Crenel', to express time or distance between 'things'. From here onwards the symbol 'C' will be used

for 'Crenel'.

'Crenels' also have various appearances to humans: 'Crenel$_D$' and 'Crenel$_T$'.

3. 'Velocity' is dimensionless number between 0 and 1 and expresses the rate of conversion between various appearances of the 'Crenel'.

4. 'Photons' are 'entities' that contain 'Packages'.

5. 'Photons' travel at the velocity of 'Photons'. The value corresponds to the dimensionless number 1.

6. Photons do not overtake each other.

7. Planck's constant ('h' in Metric Physics) is fully integrated into Crenel Physics: here it has the value of 1 PC.

8. The unit of measurement 'Package' is reciprocal to the unit of measurement 'Crenel'. Or in other words: 1 'Package' x 1 'Crenel' = 1, which is equal to the speed of light.

9. Spatial fixtures use Crenels as yardsticks along their dimension lines. Therefore, these spatial fixtures are Euclidian, and mathematical rules can be applied.

10. 'π' is a natural 0-dimensional constant. It is the outcome of a procedure: in a Crenel Physics space (= Euclidian space), divide the circumference of a circle by the diameter of that

circle.

11. Angles are expressed in 'radials'. A full revolution is represented by $2.\pi$ radials. The Radial is an absolute unit of measurement.

Force:

How about a 'unit of measurement' for force 'F'?

The physics behind formula (2.1) from Metric Physics (F = m.a) must also be valid in Crenel Physics. Here, mass is expressed in 'P'. And acceleration (m/s^2) would be expressed in 'Crenel' per 'Crenel^2', or C/C^2, which is equal to C^{-1}. Therefore, and in line with formula (2.1), 'force' is to be expressed in the unit for 'mass', multiplied with the unit for 'acceleration'. In Crenel Physics, this multiplication results in P x 1/C, which equals to P/C (= 'Packages' per 'Crenel').

Force 'F' is expressed in P/C.

The Gravitational Constant:

The gravitational constant (see chapter 2) is the *third* relevant universal physical constant: the symbol 'γ' is used. In Metric Physics, the 'gravitational constant' is:

$$\gamma = 6.670 \times 10^{-11} \text{ N.m}^2.\text{kg}^{-2}.$$

Again, as was the case for the velocity of light 'c' and for Planck's constant 'h', here is a universal constant with a numerical value (6.670×10^{-11}), which is hard to remember, and which has a unit of measurement $N.m^2.kg^{-2}$ (which also is quite complicated). And – again - these complications are root-caused by the Metric system of measurement that lies underneath.

The metric system was developed without taking into account what *true* fundamental physical properties it should reflect. In Crenel Physics this can be reconsidered, making use of what has been defined so far.

In Metric Physics, the equation for gravitational force between two objects is calculated as follows:

$$F_g = \gamma \times \frac{M_1 \times M_2}{D^2} \qquad (7.3)$$

In which:
F_g = force of gravity, in Newton
γ = gravitational constant = 6.670×10^{-11} $N.m^2.kg^{-2}$
M1 and M2 = the mass of the 'objects' at hand, in kilograms
D = the distance between the 'objects' at hand, in meters

This equation describes the strength of the attracting gravitational force between two objects, based on their metric property *mass*.

In Crenel Physics the same physical law must apply. This law must be rephrased based on the available units: the 'Package' and the 'Crenel'.

The first question to answer is: in what units does the gravitational constant 'γ' need to be expressed in Crenel

Physics? The answer can be derived by applying the appropriate units of measurement (from Crenel Physics) into formula (7.3).

To facilitate the analyses, the gravitational constant 'γ' is isolated by rephrasing formula (7.3) as follows:

$$\gamma = F_g \times \frac{D^2}{M_1 \times M_2} \tag{7.3a}$$

If one would –to explain the followed procedure called 'dimension analyses'- substitute the units of measurement from *Metric Physics* into this equation, one would get for the right part of equation (7.3a):

$$\text{Newton} \times \frac{\text{meter}^2}{\text{kg} \times \text{kg}} \tag{7.3b}$$

This is indeed equal to the unit of measurement in which the gravitational constant is expressed in Metric Physics: N.m^2.kg^{-2}.

If one performs the same dimension analyses to the right part of (7.3a), but this time using the units of 'Crenel Physics', this results in:

$$\frac{P}{C} \times \frac{C^2}{P^2} \quad \ldots$$

This can be rewritten as:

$$\frac{C}{P} \tag{7.3c}$$

Therefore, in Crenel Physics, the gravitational constant is expressed in C/P ('Crenels' per 'Package'). Its numerical value however is not yet determined. How can that be taken care of? Is this still another degree of freedom in Crenel Physics (as was the case when Planck's constant was settled to the numerical value of 1)?

To explore that question, let's review how the question was historically answered in Metric Physics: initially 'γ' did not have a numerical value yet. This was the case until equation (7.3) was discovered. At this point in time however, the units in which 'γ' was to be expressed (the Newton, the meter and the kilogram) were already quantified. So therefore, the numerical value for 'γ' became a *slave* of these units. As happened with the velocity of light ('c'), and with Planck's constant ('h').

As discussed earlier, in Crenel Physics the strategy is, to 'reverse the hierarchy' when defining the system for units of measurement. Because one now can recognize that –so far– there are *three* different and universal physical constants, these can be used as the fundaments for the new system of fundamental units of measurement. They can be put at the top of the hierarchy. This will lead to the intended normalization, as was discussed before.

When one had the choice for setting a *numerical* value for Planck's constant (through definition 5), a numerical value of 1 was selected. As a consequence, the unit of measurement 'Crenel' and the unit of measurement 'Package' became reciprocal to each other *without a numerical conversion factor* (that is: the numerical conversion factor equals 1). But although Package and Crenel are reciprocal units of measurement –as a consequence to the choice to use a conversion factor of 1– both units of measurement are still 'floating together' in terms of their *absolute* value: how much quantity one 'Package' really represents is still free… Could one lift a Package barehanded? Or does one need a hoisting crane, or is it microscopically small?

Once this quantity 'Package' would be settled in absolute terms, so would – consequently - the 'Crenel' be settled inherently… being its reciprocal.

At this point there still is a choice in Crenel Physics for the settling the numerical value of the gravitational constant. It already is recognized that the applicable unit of measurement is expressed in 'Crenels' per 'Package' (or: C/P). Thus, the choice will only impact the *numerical* value that the gravitational constant will have in 'Crenel Physics'. Again, this choice – whatever it may be - will have consequences: in this case the aforementioned 'floating together at an unknown numerical level' of the quantity of the 'Package' - and thereby the 'Crenel' - will be eliminated.

One can also review equation (7.3) to evaluate the consequence of the upcoming choice: the numerical value determines at the bottom line, how the unit of force in Crenel Physics - the left side of equation (7.3) – is going to relate to the units at the right side of equation (7.3): the 'Package' and the 'Crenel'. It is equation (7.3) that settles the relationship. And once it is settled, there will be no more degrees of freedom left for the 'Package' and/or the 'Crenel'. The numerical value of the gravitational constant effectively settles the relationship between the unit of force at the one side, and C and P at the other.

For reason of consistency it makes sense to also normalize the unit of measurement system based on this gravitational constant. Normalize it, by declaring that in Crenel Physics 'γ' equals 1 C/P. Consequently (as an example): if one takes two objects that each have a quantity of 1 'Package', and when these are positioned 1 'Crenel' apart from each other, there will be a gravitational attract force of 1 P/C (= 1 Package per Crenel).

By having *normalized* (= set to a value of 1) all aforementioned three universal physical constants…

1. The velocity of light = 1
2. Planck's constant = 1 P.C
3. Gravitational constant = 1 C/P

… one has actually used up the last degree of freedom to quantify how much a 'Package' is, or how much a 'Crenel' is, and how the unit of force is related to these. The numerical values for their units of measurement are now fixed and set by the above three physical constants.

This therefore completes –so far- a process called 'normalization' of units of measurement. But as mentioned in chapter 1, and as became clear from the above followed approach where open choices were identified (and made), there in fact are numerous choices for normalizing units of measurement. There indeed is no *physical* argument to normalize to numerical values of 1, as has been done.

The here followed choices are very similar –but not entirely similar- to a likewise normalization process that led to the so called 'Planck units'. The main justification for the slight deviation is that the here followed approach is entirely based on the current tools in the Crenel Physics toolbox, while this toolbox still is extremely basic at this point of development. There is no spin yet, nor magnetic fields or electrical forces, etcetera. There is however a strong argument to start the normalization process as early as possible in the process of developing Crenel Physics.

In the next chapter explores the consequences, expressed in metric units, of the selected Crenel Physics normalization basis.

8. Crenel Physics normalization.

Typically, as starting point of the process of 'normalization' of units of measurement, some key natural constants are presumed to have a *dimensionless* value of 1. In Crenel Physics however, this approach has not been followed. Instead, the following natural constants were defined:

1. The velocity of light = 1
2. Planck's constant = 1 P.C
3. Gravitational constant = 1 C/P

Where symbol 'C' stands for the fundamental unit 'Crenel' and 'P' stands for 'Package'. Note, that a relationship between these two fundamental units of measurement was identified (they are reciprocal to each other), and therefore these units of measurement are named 'fundamental', and not 'natural'. This is because the relationship suggests an even more basic layer underneath the Crenel and the Package.

Only the velocity of light was defined as a dimensionless natural constant. The Crenel and the Package both appear in the unit of measurement of Planck's constant, as well as in the gravitational constant.

In the previous chapter the hierarchy in units of measurement was thus set. Thereby, the three listed universal physical constants were placed at top level, and three units of measurement in Crenel Physics - P (for 'Packages'), C (for 'Crenels') and the 'Unit of Force' (P/C) - were based on these. They thus became *slaves* of these three universal physical constants. Their numerical value is now *absolute*, because the value of their masters was –arbitrarily- set to a numerical value of 1, and because these masters themselves are absolute: the set numerical values are –presumed- the same for anyone, anywhere, within any frame of reference.

As discussed, the term 'absolute' means that one could give instructions by phone to determine their values. In Crenel Physics one would have to give three instructions 'by phone' to ensure compatibility in using the aforementioned Crenel Physics units of measurement. These three instructions are:

1. How to determine the velocity of Photons through instructions by phone.

   Actually: there would be no complicated instructions here… the velocity of Photons *is* the velocity of Photons… whatever the circumstances may be. The instruction by phone would be: this velocity is 1 (one), and dimensionless.

   Note that in Metric Physics the velocity of light is *not* constant: here, this velocity (expressed in meters per second) can be slowed down by a medium through which Photons are traveling. This slowdown is caused by Photon scattering, as was previously discussed. The velocity of light is only universally constant when measured in a vacuum. The propagating of Photons through a medium (like glass) has not yet been addressed in

Crenel Physics.

2. Instruct that Planck's law is valid, and that Planck's constant equals 1 P.C (one 'Package' times one 'Crenel').

   This has the consequence that 1 Package corresponds to a quantity that one would find contained in an object that has a frequency of 1 revolution per Crenel. And instruct that -because of Planck's law- the 'Package' and the 'Crenel' actually are *reciprocal* units of measurement.

3. Instruct that if one takes two of these objects with a frequency of 1 revolution per Crenel, and position these one 'Crenel' apart, the resulting gravitational - and also mutual - attract force equals 1 Package per Crenel.

   That's the definition for the unit of force, which is based here on gravitational interaction.

Because the laws of physics do as well apply in Crenel Physics as they do in Metric Physics, it now must be possible to find conversion factors between the units that are used in Crenel Physics (being 'Crenel', 'Package' and 'Unit of Force'), and the units of Metric Physics (being Joule, kilogram, meter, second and Newton). And these conversion rates -one can logically assume- are solely depending on the three given physical constants. There *could* be no other dependencies because there *was* nothing else required to fix these units in Crenel Physics... The Package and the Crenel thus indeed are the children of the three mentioned universal physical constants, with a clear bloodline (so to speak).

Prior to finding these conversion factors, it is to be recognized that it is the *physical* reality of these universal constants that makes such a conversion possible. In both Metric Physics and in Crenel Physics, the velocity of light 'c' is not just a velocity: it has a special meaning and relevance. Like Plank's constant 'h', and like the gravitational constant 'γ'. No matter where one is or what one is doing (or how fast one is travelling): these three physical constants are presumed to *keep their constant values*. In Crenel Physics all three received a numerical value of 1.

One can also look at the aforementioned presumption from a different angle: should one of the natural constants in fact float around in value (through the eyes of some objective observer), so would the Package, the Crenel and the Unit of Force float around consistently, such that physical laws based on these are not impacted.

The subscript 'new' is used to differentiate units in Crenel Physics from units used in Metric Physics. For these three universal physical constants one can now summarize:

- $c_{new} = c$ (velocity of light)   (8.1a)
- $h_{new} = h$ (Planck's constant)   (8.1b)
- $\gamma_{new} = \gamma$ (gravitational constant)   (8.1c)

The above three equations are true when the *new* versions of physical constants are expressed in the units of Crenel

Physics, while the *old* versions are expressed in the units of Metric Physics.

It is through these three equations, that all looked for conversion rates between Metric Physics and Crenel Physics can be found.

As a first step, equation (8.1a) can be rewritten, taking the respective units of measurement into account. The left term of the equation 'c_{new}' is a dimensionless number with value 1. The right term 'c' is expressed in m/s second. Equation (8.1a) can therefore be written as:

$$1 = c \text{ (m/s)} \tag{8.2}$$

…where the numerical value of 'c' is approximately 2.9979×10^8.

Equation (8.2) can be rewritten as:
1 second = c meter. This confirms that the parameter 'time' (in seconds) can indeed be converted into 'distance' (in meters): one second is the equivalence of approximately 2.9979×10^8 meters.

Likewise, equation (8.1b) can be rewritten as:

$$1 \text{ (P.C)} = h \text{ (J.s)} = h \text{ (N.m.s)} \tag{8.3}$$

…where the numerical value of 'h' is approximately 6.6256×10^{-34}.

Note, that in the metric system one Joule (J) is equivalent to one N.m (=Newton.meter).

And likewise, equation (8.1c) results in:

$$1 \ (C/P) = \gamma \ (N.m^2.kg^{-2}) \tag{8.4}$$

…where the numerical value of 'γ' is approximately 6.670×10^{-11}.

In recognizing that the units Newton, kilogram, meter, second of the Metric Physics are set and known, and that - based on these - the three physical constants 'c', 'h' and 'γ' received their numerical values, equation (8.3) and (8.4) represent two equations with two *unknown* parameters (C and P). From these two equations, the two unknown parameters 'C' and 'P' can be resolved through simple mathematics, as follows:

Step 1:
Isolate P from equation (8.3):

$$1 \ P = h \ N.m.s.C^{-1} \tag{8.5}$$

Step 2:
Substitute this value for P in equation (8.4):

$$1 \ C = \gamma.h \ \ N.m^2.kg^{-2} \ . \ N.m.s.C^{-1} \tag{8.6}$$

In equation (8.6), multiply left and right term with factor 'C' ('Crenel'):

$$C^2 = \gamma.h \ \ N^2.m^3.kg^{-2}.s \tag{8.7}$$

Step 3:

Because 1 kg is the equivalent of c^2 (Joules)… according to $E = mc^2$… 1 kg is equivalent to: $1 \text{ kg} = 1 \text{ J}.c^2 = 1 \text{ N.m}.c^2$.

The 'kg^{-2}' (as found in equation (8.7)) thus corresponds to the alternative notation: '$N^{-2}.m^{-2}.c^{-4}$'.

And because 1 second is the equivalent of c meters (per equation 8.2), substitute in equation (8.7) for variable 's' the alternative notation:
'c.m'.

Both substitutions combined result in:

$$C^2 = \gamma.h \text{ N2.m3. } N^{-2}.m^{-2}.c^{-4}.c.m$$
$$= \gamma.h.c^{-3} \text{ } m^2 \tag{8.8}$$

Step 4:
Take square root:

$$\textbf{1 Crenel} = \sqrt{\frac{\gamma.h}{c^3}} \textbf{ meter} \tag{8.9}$$
$$= \pm 4.05 \times 10^{-35} \text{ meter}$$

And, because 1 meter = 1 second/c:

$$\textbf{1 Crenel} = \sqrt{\frac{\gamma.h}{c^5}} \textbf{ seconds} \tag{8.10}$$
$$= \pm 1.35 \times 10^{-43} \text{ seconds}$$

Step 5:

Substitute (8.9) in (8.5):

$$P = h \ N.m.s. \ C^{-1} = h \ N.m.s. \ \sqrt{\frac{c^3}{\gamma.h}} \ m^{-1}$$

$$= \sqrt{\frac{c^3.h}{\gamma}} \ N.s \tag{8.11}$$

Because 1 second is the equivalent of c meters:

$$P = \sqrt{\frac{c^3.h}{\gamma}} \ N.s = \sqrt{\frac{c^3.h}{\gamma}} \ N.c.m = \sqrt{\frac{c^5.h}{\gamma}} \ N.m$$

$$= \sqrt{\frac{c^5.h}{\gamma}} \ \text{Joule} \tag{8.12}$$

Or:

$$\textbf{1 Package} = \sqrt{\frac{c^5.h}{\gamma}} \ \textbf{Joule} \tag{8.13}$$
$$= \pm 4.90 \times 10^9 \ \text{Joule}$$

Or:

$$\textbf{1 Package} = \sqrt{\frac{c.h}{\gamma}} \ \textbf{kilogram} \tag{8.14}$$
$$= \pm 5.46 \times 10^{-8} \ \text{kilogram}$$

And, for the reversed conversions:

$$\textbf{1 second} = \sqrt{\frac{c^5}{\gamma.h}} \ \textbf{Crenel} \tag{8.15}$$
$$= \pm 7.40 \times 10^{42} \ \text{Crenel}$$

1 meter $= \sqrt{\dfrac{c^3}{\gamma.h}}$ **Crenel** $\hspace{2cm}$ (8.16)

$= \pm\, 2.47 \times 10^{34}$ Crenel

1 Joule $= \sqrt{\dfrac{\gamma}{c^5.h}}$ **Package** $\hspace{2cm}$ (8.17)

$= \pm\, 2.04 \times 10^{-10}$ Package

1 Kilogram $= \sqrt{\dfrac{\gamma}{c.h}}$ **Package** $\hspace{1.5cm}$ (8.18)

$= \pm\, 1.83 \times 10^{7}$ Package

From the above, the unit for *force* in Crenel Physics can be derived, because in Metric Physics 1 Newton equals 1 Joule/meter while in Crenel Physics this unit of force is expressed in P/C.

> *Note: from here onwards, the symbol 'UoF' will be used as the unit of force in Crenel Physics.*

To achieve this, one can use equation (8.13) that converts Packages to Joules, and formula (8.9) that converts Crenels to meters. Dividing formula (8.13) by formula (8.9) gives – expressed in Joule/meter – the conversion rate between one UoF in Crenel Physics, and the Newton:

$$1\ \text{UoF} = \frac{1\ \text{Package}}{1\ \text{Crenel}} = \frac{\sqrt{\dfrac{c^5.h}{\gamma}}}{\sqrt{\dfrac{\gamma.h}{c^3}}}\ \frac{\text{Joule}}{\text{meter}}$$

1 UoF = $\dfrac{c^4}{\gamma}$ **Newton** (8.19)

$= \pm 1.21 \times 10^{44}$ Newton

And for the reversed conversion:

1 Newton = $\dfrac{\gamma}{c^4}$ **UoF** (8.20)

$= \pm 8.26 \times 10^{-45}$ UoF

All the above results are the conversion factors between fundamental physical units in Crenel Physics (the 'Package' and 'Crenel') and the *appearances* of these fundamental units in Metric Physics. These are subjective appearances to human beings. Because the 'Crenel' has *two* appearances to humans (namely time and distance), there also are two associated conversion factors: equations (8.9) and (8.10) respectively. Perhaps, in the future more appearances may be defined. For each of these, a conversion factor would need to be identified.

Likewise, for the 'Package' three appearances were discussed: mass, energy and frequency. The conversion factors for the first two appearances are given by equations (8.13) and (8.14) respectively. The still missing third conversion factor - between the 'Package' and the appearance 'frequency' - can be derived as follows:

Based on the equation $E = h.\upsilon$ (or: $\upsilon = E/h$) the conversion factor from 'Package' to 'Joules' - as given by equation (8.13) - needs to be divided by 'h'. Because in Metric Physics, frequency is expressed in the unit Hertz (symbol Hz, revolutions per second), this results in:

$$1 \text{ Package} = \sqrt{\frac{c^5 . h}{\gamma}} \times \frac{1}{h} \text{Hertz}$$

$$\mathbf{1 \text{ Package}} = \sqrt{\frac{c^5}{\gamma . h}} \text{ Hz} \qquad (8.21)$$

$$= \pm 7.4 \times 10^{42} \text{ Hz}$$

Again it is thinkable, that more appearances of the Package are identified in the future. For example 'number of degrees of freedom' or 'information content'. In any of these cases, a conversion factor towards 'Packages' is to be defined.

With for all current appearances the conversion factors in place, it now is possible to further elaborate on the earlier conclusion that in Crenel Physics the 'Package' ('P') and the 'Crenel' ('C') are *reciprocal units of measurement*. This reciprocal relationship can mathematically be expressed by the equation:

$\mathbf{P.C=1}$ (dimensionless) $\qquad (8.22)$

This further elaboration addresses the conversion factors that are associated with the various *appearances* that the Package and the Crenel can have (to humans). Because – as discussed so far - the Package has *three* appearances and the Crenel has *two* appearances to humans, there would be *six* possible combinations for addressing equation (8.22) in Metric Physics.

That is: six combinations for multiplying the Package conversion factor with a Crenel conversion factor. The table below shows the result of these 6 combinations, where each cell shows the product of the two associated conversion factors (per equation 8.22):

	P_E Energy appearance	P_M Mass appearance	P_F Frequency appearance
C_t Time	$\sqrt{\dfrac{\gamma.h}{c^5}}$ $\times \sqrt{\dfrac{c^5.h}{\gamma}}$	$\sqrt{\dfrac{\gamma.h}{c^5}}$ $\times \sqrt{\dfrac{c.h}{\gamma}}$	$\sqrt{\dfrac{\gamma.h}{c^5}}$ $\times \sqrt{\dfrac{c^5}{\gamma.h}}$
C_d Distance	$\sqrt{\dfrac{\gamma.h}{c^3}}$ $\times \sqrt{\dfrac{c^5.h}{\gamma}}$	$\sqrt{\dfrac{\gamma.h}{c^3}}$ $\times \sqrt{\dfrac{c.h}{\gamma}}$	$\sqrt{\dfrac{\gamma.h}{c^3}}$ $\times \sqrt{\dfrac{c^5}{\gamma.h}}$

Table 8.1: six possibilities for multiplying P with the C, based on various appearances (to humans).

The above table (8.1) can be mathematically rewritten - and thereby simplified - as follows:

	P_E Energy	P_M Mass	P_F Frequency
C_T Time	'h'	'h/c^2'	'1'
C_D Distance	'c.h'	'h/c'	'c'

Table 8.2: a clone of table (8.1)... but mathematically simplified.

Table (8.2) shows a remarkable symmetry. This symmetry becomes apparent if one 'hops' between the various appearances of the Package (the 3 options in the columns of the table) and/or hops between the various appearances of the Crenel (the 2 options in the rows of the table):

1. if one 'hops' rows (from the 'Time' appearance of the Crenel 'C_T' to its 'distance' appearance 'C_D'), the cells in the table show that – to get the value of the 'distance' appearance - this requires a multiplication with a factor 'c' (the velocity of light): all terms in the C_D-row are equal to the terms in the C_t-row, multiplied by the factor 'c'. And hopping back would require a division with 'c'.

    The table thus shows that – *irrespective of the column in the table (or: which of the three listed Package appearances one selects)* – the conversion that is required for hopping between the two Crenel appearances is *not* impacted by this choice.

    This would lead to the following hypothesis: should an additional (here: 4th, 5th, etc) appearance of the Package be proposed, the same rule would apply: such new Package appearance would pop up in all possibilities of the Crenel appearance (until here there are only two, namely 'distance' and 'time'), and between these two Crenel appearances this same factor 'c' would be expected.

One potential new candidate for the Package appearance could be the 'bits of information'. Another candidate could be '(the number of) degrees of freedom'. Each such new appearance would add a new column in tables (8.1) and (8.2), and the respective conversion factors in the rows within that column would need to be determined. But one would expect –again (and according to the aforementioned hypothesis)- a factor 'c' between the two shown rows.

2. 'Hopping' between the three shown appearances of the Package (Energy, Mass and Frequency) also shows consistent conversions. Between the three shown columns there are three possible combinations of hopping back and forth:

   a. The 'hopping' back and forth between the *Energy* and the *Mass* appearance of the Package involves respectively dividing or multiplication with a factor c^2.

   This - of course - is in effect the representation of Einstein's equation $E=m.c^2$ (in Metric Physics), which describes the transformation between these two appearances.

   The table again shows that this 'Package appearance transformation' procedure does *not* depend on the selected row (= the selected Crenel appearance).

b. The 'hopping' back and forth between the *Energy* and *Frequency* appearance of the Package involves division or multiplication with a factor 'h'.

This is in effect the representation of Planck's equation E=h.υ, which describes the transformation between these two appearances in Metric Physics.

And again, the selection of the row in the table (= the selected Crenel appearance) has no impact in this Package conversion.

c. The 'hopping' back and forth between the *Mass* and the *Frequency* appearance of the Package involves division or multiplication with a factor h/c^2 respectively.

Again, this factor is not impacted by the choice of row in the table (= the selected Crenel appearance).

Table (8.2) also shows that the gravitational constant 'γ' plays *no role at all* in the conversion between the listed appearances. The gravitational constant 'γ' does not appear in the table.

Apparently the gravitational constant only plays a role in conversion back and forth between the 'Crenel Physics' units of measurement, and their counterpart units of measurement in Metric Physics. But the gravitational constant -at the bottom line- plays *no role at all* in the conversion between appearances in table (8.2).

> Because the role of the gravitational constant -as one separate natural constant- is irrelevant to the above shown 'appearances', it apparently plays a *more basic* role in Crenel Physics, closer to true *natural* units. Apparently, the gravitational constant plays a role *underneath* the level of 'appearances' in Crenel Physics. Note that this same level (of what was called 'appearances' in the context of Crenel Physics) relates to 'fundamental units' in Metric Physics...
> Thus, gravity does not play a role as a 'fundamental constant' anymore in Metric Physics: it is even *more* basic than 'c' and 'h'. Also based on what was discussed before, gravity can therefore best be redefined as:
>
> *Gravity:*
> *'one possibility of interaction between pairs of individual objects, perhaps even <u>the</u> possibility of interaction'.*
>
> This is a very dramatic definition if one realizes that without any form of interaction between objects our entire world would fall apart. As will be shown later, gravity indeed plays this dramatic role in Crenel Physics.

The number '1' in the upper right cell of table (8.2) expresses that the associated appearances seem the *base case* in relation to Crenel Physics: when the 'Crenel' is associated with its 'Time' appearance (to humans), and simultaneously the 'Package' is associated with its 'Frequency' appearance (to humans), no special factor is required to reflect that the 'Package' is indeed the reciprocal of the 'Crenel'. The conversion factor for the product P.C equals the dimensionless value 1, in this particular case.

However, as soon as combinations with other 'appearances' come into the picture, in Metric Physics the additional factors - 'c' (the speed of light) and 'h' (Planck's constant) – are required for conversion, per table (8.2).

For completeness, table (8.3) below shows what the normalization in Crenel Physics produced: here Packages are expressed in Packages, and Crenel in Crenel. Therefore there is *no* conversion between the various shown appearances. Or: all conversion factors between appearances are equal to the dimensionless value '1'.

	P_E (Energy)	P_M (Mass)	P_F (Frequency)
C_T (Time)	'1'	'1'	'1'
C_D (Distance)	'1'	'1'	'1'

Table 8.3: the Crenel Physics version of table (8.2).

9. Verification.

The highlighted equations in the previous chapter show that the conversion factors between Metric Physics units of measurement and the Crenel Physics units of measurement are entirely based on three physical constants: 'c', 'h' and 'γ'.

Given the relevance of these conversion factors (they form the mooring lines of Crenel Physics towards Metric Physics), this chapter *verifies* that these factors are indeed consistent with three earlier used physical equations. Furthermore, two additional equations -that will be used in the following chapters of this book– will be verified for consistency with these conversion factors and for further usage in Crenel Physics: the equations for 'gravity force' and for 'centripetal force'.

Therefore, this chapter adds five well known equations to the Crenel Physics toolbox. Other than verification for consistency thereof, this chapter adds no new insights into Crenel Physics.

Obviously, physical laws in Metric Physics also are valid in Crenel Physics. The fact that different units of measurement are used in Crenel Physics should not undermine these physical laws.

1. Verification of consistency with Crenel Physics of Einstein's equation: '$E = m.c^2$'.

   This equation was used in finding the conversion factors (see step 3 in the previous chapter).

If this law is applied to 1 kg, it would read:
1 Joule = 1 kilogram * c^2

From this, 'c' can be extracted:

$$c \left(\frac{m}{s} \right) = \sqrt{\frac{1 \text{ Joule}}{1 \text{ kilogram}}}$$

(in Metric units)

Using equation (8.17), the variable '1 Joule' can be converted into the unit 'Package'. Likewise, equation (8.18) converts the variable '1 kilogram' into the unit 'Package'. Thus, the entire *right* term of the above equation can be converted into the units of Crenel Physics.

To keep the equation valid after the conversion, the *left* term also must be converted to units of Crenel Physics. Where in Metric Physics 'c' is equal to 'c' in m/s, the value of 'c' in Crenel Physics is -by definition- the dimensionless number of 1. To achieve the conversion -from what the left term 'c' *is*- to what it *should* become (a dimensionless number with value 1), one needs to convert the unit of measurement m/s using equations (8.16) and (8.15) respectively, resulting in a factor 1/c:

$$\frac{m}{s} = \frac{\sqrt{\frac{c^3}{\gamma.h}}}{\sqrt{\frac{c^5}{\gamma.h}}} = \frac{1}{c}$$

The above equation would thus translate into Metric Physics to:

$$1 \times \frac{1}{c} = \sqrt{\frac{1 \text{ Joule (converted)}}{1 \text{ Kilogram(converted)}}}$$

Or:

$$1 = c \times \sqrt{\frac{1 \text{ Joule (converted)}}{1 \text{ Kilogram(converted)}}}$$

Which after using equations (8.17) and (8.18) for the conversions results in:

$$1 = c \times \sqrt{\frac{\sqrt{\frac{\gamma}{c^5.h}}}{\sqrt{\frac{\gamma}{c.h}}}} =$$

$$= \sqrt{\sqrt{\frac{\gamma}{c^5.h}} \times \sqrt{\frac{c.h}{\gamma}}} = c \times \sqrt{\sqrt{\frac{1}{c^4}}} = c \times \frac{1}{c} = 1$$

Because in the above equation left side and right side are indeed equal, the formula $E = m.c^2$ is verified in Crenel Physics.

2. Verification of consistency with Crenel Physics of Planck's equation: 'E=h.υ'.

This equation was used to find the conversion factors (see Equation (8.21)).

If this law is applied to an object that has a frequency 'υ' of 1 cycle per second, the associated energy would be 'h' Joules.

Thus,

$h = E/υ$

where both the left term as well as the right term are expressed in J.s in Metric Physics.

Again, if both sides of the equation are converted into units of Crenel Physics, the left term should result into value 1. To achieve the over-all conversion, the left term 'h' needs to be divided by 'h'. That operation is effectively the same as multiplying the right term with 'h'. The result of the conversion into units of Crenel Physics would be as follows:

$$1 = h . 1\, Joule(converted) . 1\, second(converted)$$

To verify this, equations (8.17) and (8.15) can be used respectively:

$$1 = h \times \sqrt{\frac{\gamma}{c^5.h}} \times \sqrt{\frac{c^5}{\gamma.h}} = h \times \sqrt{\frac{1}{h^2}} = 1$$

Because in the above equation left side and right side are indeed equal, the formula $E = h.υ$ is verified to be applicable in Crenel Physics.

3. Verification of consistency with Crenel Physics of Newton's equation: 'F=m.a'.

The left term in this equation can be converted into Crenel Physics units using equation (8.20):

$$\frac{\gamma}{c^4}$$

The right term is expressed in $kg.m.s^{-2}$, which can be converted using equations (8.18), (8.16) and (8.15) respectively:

$$\sqrt{\frac{\gamma}{c.h}} \times \sqrt{\frac{c^3}{\gamma.h}} \times \frac{\gamma.h}{c^5} = \frac{\gamma}{c^4}$$

Because the right term of the equation converts to the same result as the left term, the formula $F = m.a$ is verified.

4. Verification of consistency with Crenel Physics of equation for gravitational force: '$F_g = \gamma.M_1.M_2/D^2$'.

The force of gravity F_g will – after the conversion into Crenel Physics units of measurement - be expressed in P/C.

γ, the gravitational constant, will – after the conversion - be expressed in C/P.
Masses M1 and M2 will – after the conversion - both be expressed in P.

Distance D will be converted to C.

Thus, the right term of the above equation will be expressed in:

$$\frac{C}{P} \times P^2 \div C^2 = \frac{P}{C}$$

This is the same dimension as the left term. Thus, from a dimensional analyses point of view the equation for the force of gravity is verified.

To also verify the respective conversion factors: the conversion factor that will be applied to the left side of the equation must be equal to the – over all - conversion factor that will be applied to the right side of the equation.

The conversion factor that will be applied to the left side of the equation is given by equation (8.20), as the force F_g is expressed in Newton:

$$\frac{\gamma}{c^4}$$

The parameters at the right side of the equation can be converted using equations (8.18) and (8.16) respectively, and by taking into account that the gravitational constant γ will convert to value 1, which is reflected by dividing the right term by γ:

$$\frac{1}{\gamma} \times \frac{\gamma}{c.h} \times \frac{\gamma.h}{c^3} = \frac{\gamma}{c^4}$$

Thus, the conversion factor applied to the left term of the equation is the same, as the conversion factor that is applied to the entire right term of the equation. Thereby the equation to calculate the force of gravity has been verified.

5. Verification of consistency with Crenel Physics of centripetal force: '$F_{cp} = M.\omega^2.r$'.

The centripetal force F_{cp} will - after the conversion into Crenel Physics units of measurement - be expressed in P/C.

Mass M will – after the conversion - be expressed in P.
The rotation speed ω will be expressed in radials/C.
Radius 'r' will be converted to C.

The – over all - right term of the equation will thus be expressed in:

$$P \times \frac{1}{C^2} \times C = \frac{P}{C}$$

This is the same dimension as the left term. Thus, from a dimensional analyses point of view the equation is verified.

To also verify the respective conversion factors: the conversion factor that will be applied to the left side of the equation must be equal to the – over all - conversion factor that will be applied to the right side of the equation.

The conversion factor that will be applied to the left side of the equation is given by equation (8.20), as the force F_{cp} is expressed in Newton:

$$\frac{\gamma}{c^4}$$

The parameters at the right side of the equation can be converted using
Equations (8.18), (8.15) and (8.16) respectively. The over-all conversion factor thus becomes:

$$\sqrt{\frac{\gamma}{c.h}} \times \frac{\gamma.h}{c^5} \times \sqrt{\frac{c^3}{\gamma.h}} = \frac{\gamma}{c^4}$$

Thus, the – over all - conversion factor for the right side of the equation matches the conversion factor for the left side of the equation, and therefore the equation is verified.

With the above five equations at hand, some basic mechanics are now possible in Crenel Physics.

10. Cruising through Crenel Physics.

This chapter cruises through some viewpoints from the Crenel Physics perspective. It forms a starting point for the question: what's next? Now that two fundamental – normalized- units of measurement have been defined (the Package and the Crenel), how can these be used to construct the reality around us?

In summary, these are some viewpoints:

1. One can only detect an object if it contains a non-zero number of 'Packages'.

   The 'Package' (symbol: 'P') replaces the metric units of measurement for mass (kg), energy (J) and frequency (Hz). The latter are to be considered as *appearances* (to humans) of the 'Package'.

   The number of 'Packages' that an object contains is the same as 'the Strength of Gravity' that comes with the object. This 'Strength of Gravity' only appears between *pairs* of objects as an exclusive relationship.

2. To humans, objects appear to reside in a 'space' (a spatial Fixture) that can be defined using e.g. spherical coordinates.

   In defining a 'space' to humans, there are (at least) two *types* of coordinates or dimensions one may require:

- Angles,
  expressed in 'radials' relative to a direction of reference.

  The 'radial' is an absolute (mathematically defined) unit of measurement.

- 'Crenels',
  measured from some reference point.

  The 'Crenel' (Symbol 'C') replaces the metric units for distance (m) and time (s). The latter are considered to be *appearances* (to humans) of the Crenel.

3. 'Packages' and 'Crenels' are reciprocal units of measurement:

   - If an object contains 25 'Packages', this is equivalent to containing 25 'Crenels^{-1}'.

   - Or, if two objects are 30 'Crenels' separated from each other, this is equivalent to being separated 30 'Packages^{-1}' from each other.

4. 'Force' is expressed in P/C.

The following are the three universal physical constants that have been addressed so far, and that form the basis for the units of measurement in Crenel Physics:

1. The velocity of light: 'c' = 1
2. Planck's constant 'h' = 1 P.C
3. Gravitational constant 'γ' = 1 C/P

Besides these building blocks and universal physical constants, several physical understandings have been explored:

1. An object can only be detected if 'something moves'.

   In Metric Physics one would state that an object must exist for at least some period of 'time' in order to be detectable. The latter formulation assumes that there is an understanding what 'time' really is… and at least for so far there is no understanding.

   The 'Crenel Physics' formulation at the other hand requires an alternative understanding: an understanding of 'movement' (or 'velocity')…

   In Crenel Physics, 'movement' is expressed as a *dimensionless* number between 0 and 1. Movement expresses a *conversion* between two appearances (time and distance) of the Crenel.

   There is not much one could possibly understand as it comes to the *dimension* of movement:

movement is *dimensionless*.

2. E=m.c2=h.υ

This equation expresses physical facts that are
fully integrated into -and formed the basis of-
Crenel Physics. Therefore –as an equation- it has
become obsolete. This obsoleteness is the logical
outcome of the units of measurement in Crenel
Physics, being based on the aforementioned 3
physical constants.

For all three terms in the above equation, only
one single new unit of measurement has been
introduced: the 'Package'.

Thus, the impact of the given equation is included
in the definition of the Package.

3. F=m.a

In Crenel Physics the associated units of
measurement are:
   a. 'F' is expressed in P/C
   b. 'm' is expressed in P
   c. 'a' is expressed in C/C^2 (= C^{-1})
Consequently – because of the definition of the
Crenel and the Package being based on
fundamental physical units – relativistic effects
are inherently taken into account here.

Or: in Crenel Physics the theory of relativity is
embedded into the units in which this equation is
expressed. The equation is valid (= holds) when

Crenel Physics units of measurement are used. Or; the numerical value of 'mass' –when expressed in Packages- is not relativistic.

4. $F_g = \gamma . M_1 . M_2 / D^2$

F_g is the exclusive gravity attract force between a pair of objects, which is based on the 'Strength of Gravity' (= 'Packages') that each individual object contains. In Crenel Physics the associated units of measurement are:

    a. 'F_g' in P/C
    b. 'γ' in C/P
    c. 'M_1', 'M_2' in P, which is equal to the 'Strength of Gravity' of each object
    d. 'D' in C

Again, using these Crenel Physics units of measurement, the equation inherently covers relativistic effects.

5. $F_{cp} = M . \omega^2 . r$

In Metric Physics the associated units of measurement are:

    a. 'F_{cp}' in P/C
    b. 'M' in P
    c. 'ω' in radials/C
    d. 'r' (radius) in C

And again this equation will hold under relativistic circumstances, provided that the Crenel Physics units of measurement are applied.

11. Planck's constant = the velocity of light.

As was discussed, 'Packages' and 'Crenels' are mutually reciprocal units of measurement. Therefore, in Crenel Physics these units of measurement cannot be seen as independent anymore. This chapter discusses the impact of the relationship.

In general, any physical parameter is expressed in a *numerical value* followed by its *unit of measurement*. For example:

- the ambient temperature is 10.6 degrees Celsius,
- the volume of this beer glass is 0.48 liter,
- a particle contains 43.6 Packages,
- your distance is 26.2 Crenels away from me,
- 20 Crenels ago (... I stopped the car),
- etc.

In Crenel Physics, the parameter 'velocity' is a special case: it is *dimensionless*, and may vary between 0 and 1.The velocity of light is 1.

Now consider the following:

> To retrieve the reciprocal value of any of the above parameters, both the numerical value of the original parameter, as well as its units of measurement are shifted into the denominator of a division. Thus, the reciprocal of 'X' is '1/X', whatever 'X' may be.

However, the situation explored here deviates from the above: the Package being the reciprocal of the Crenel is *not* involving the *numerical* part of the parameter. Only the *unit of measurement* is under investigation.

To clarify this, review the following statements:

- 'You are 25 Crenels away from me'
  is equal to:
  'You are 25 Packages^{-1} away from me'.

  or:

- 'This stone has a quantity of 36 Packages'
  is equal to:
  'This stone has a quantity of 36 Crenels^{-1}'.

The *numerical* value does not change when the alternative (and reciprocal) unit of measurement is used.

This puts Planck's constant into another perspective: in Crenel Physics, Planck's constant 'h' equals 1 C.P. Thus, one can now conclude that Planck's constant – in Crenel Physics – is in effect *dimensionless*: the dimension of C.P can be rewritten as $C.C^{-1}$ (or $P^{-1}.P$), which would then make it equal to a dimensionless number with a numerical value of 1.

> Consequently:

> *in Crenel Physics, Planck's constant 'h' is de facto **the same** universal fundamental physical property (and thus is equal to...) the velocity of*

light 'c': in Crenel Physics this velocity 'c' is also equal to the dimensionless number 1.

The above is however not covering the full spectrum, because in Crenel Physics there *are* two separate physical units of measurement ('C' and 'P'), albeit that these two are related through the equation C.P = 1.

To investigate the consequence of the existence of this relationship, consider the following graph (Figure 11.1) of the equation Y = 1/X.

This graph gives a 'broader' view than just stating that 1 Crenel = 1 / Package, or that 1 C = 1 P^{-1}.

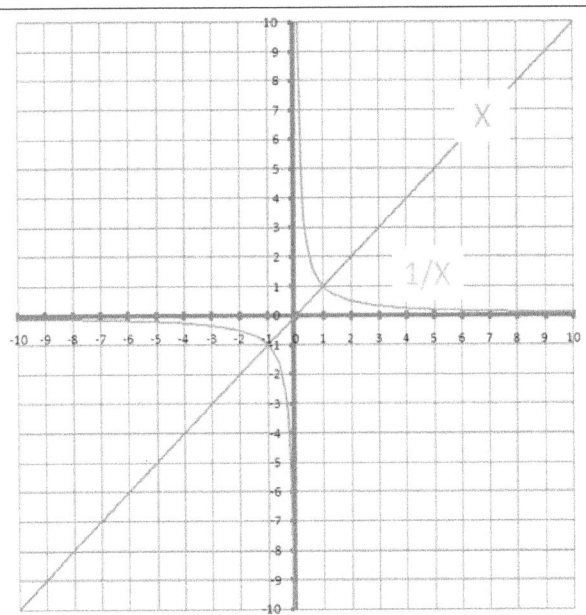

Figure 11.1: 'X' and its reciprocal '1/X'.

Because 'P' and 'C' are reciprocal units of measurement, Planck's constant is -at closer look- not fully reflected by a singular dimensionless mathematical point, with value 1 (as is the speed of light 'c'). Instead, Planck's constant is actually 2-dimensional in terms of its units of measurement, and it is calculated as the product of the line marked 'X' in figure (11.1) and the line marked '1/X' in figure (11.1). This is so for any arbitrarily chosen value of 'X'.

And obviously the result of the product equals 1 ... But there is *one* exceptional point in this figure: the point where X equals the value 0. For this point one would need to multiply the value '0' (for 'X') with an infinite high (or low) value for 1/X. This – in mathematics – causes trouble: the outcome of such a multiplication is undefined.

The physical consequence is that Planck's constant *cannot* be applied to cases where either zero Packages or zero Crenels are involved (note, that figure 11.1 is entirely symmetrical and that 'X' can stand for either the 'Package' or as well the 'Crenel').

In Crenel Physics this leads to the following interpretations:

- The conversion into a reciprocal parameter has been tackled: the unit 'Crenel' is equal to the unit 'Package^{-1}', and the unit 'Package' is equal to 'Crenel^{-1}' provided that the numerical value of the parameter is *not equal to 0* (zero).

o In these 'non-zero' cases, in Crenel Physics, Planck's constant is de facto equal to the speed of light 'c', both from a numerical point of view as well as from a unit-of-measurement point of view: both are *dimensionless* physical constants with a numerical value of 1.

o In the 'zero case' however, in Crenel Physics, Planck's constant cannot be applied.

In other words: *there must be something around* (Packages or Crenels), before Planck's constant 'h' becomes available, apparent and applicable.

Whereas the value for the velocity of light initially has been assumed to be available under *all* circumstances as a universal physical constant -so even if there is absolutely nothing around- this assumption now must be fine-tuned as well: there must be 'something' around for the 'c' to become available, apparent and applicable.

• The physical meaning of both a 'Package' as well as a 'Crenel' has been addressed in Crenel Physics:

they are the Yin and the Yang.

Albeit that in Crenel Physics they are not *counter poles* to each other - as the old theory says - but *reciprocals* to each other.

- ○ **'Packages' offer the possibility to express some substance, some quantity.**
- ○ **'Crenel' offer the possibility to express the residing somewhere, sometime.**
- ○ **And they <u>always</u> come together.**

This dual requirement -to allow detection of an object- (which was: that it should contain substance *and* exists for some period of time) therefore is at its bottom line a *single* requirement:

> *If* 'an object' contains Packages, 'this object' *does* exist for some period of time.

12. A basic model

Figure (12.1) expresses - in Crenel Physics terminology - a model for a human's subjective perception of an object. At the top of the figure is 'Me', representing a subjective individual human observer. And at the bottom is any- 'thing' representing some object.

In between are three layers:

1. 'Sensors'.

   This layer generates subjective input signals that arrive into the human brain (indicated as 'Me'). These signals are delivered by sense organs to the human brains. Sometimes these organs are extended through instrumentation.

   A separation has been made between 'Virtual Sensors' and 'Physical Sensors'. The first type of sensors relates to human perceptions that are related to intangible sensing, such as the perceptions of 'time' and 'distance'.

2. 'Appearances'.

   This layer shows how a variety of signals that can potentially be sensed (see 1 above) based on one single underlying property (see 3 below).

   In Metric Physics this layer of appearances is – other than in Crenel Physics- referred to as

'Fundamental Physical Units'.

3. 'Fundamental Physical Units'.

    This is a new layer... typical for Crenel Physics and coming forth from the here followed 'normalization' process.

    Note, that this layer still does not represent pure 'natural units' because there is a found relationship between Crenels and Packages (see previous chapters). In fact, one could consider the any-'thing' at the bottom of the figure as *the* natural unit, as it produces both Crenels and Packages simultaneously.

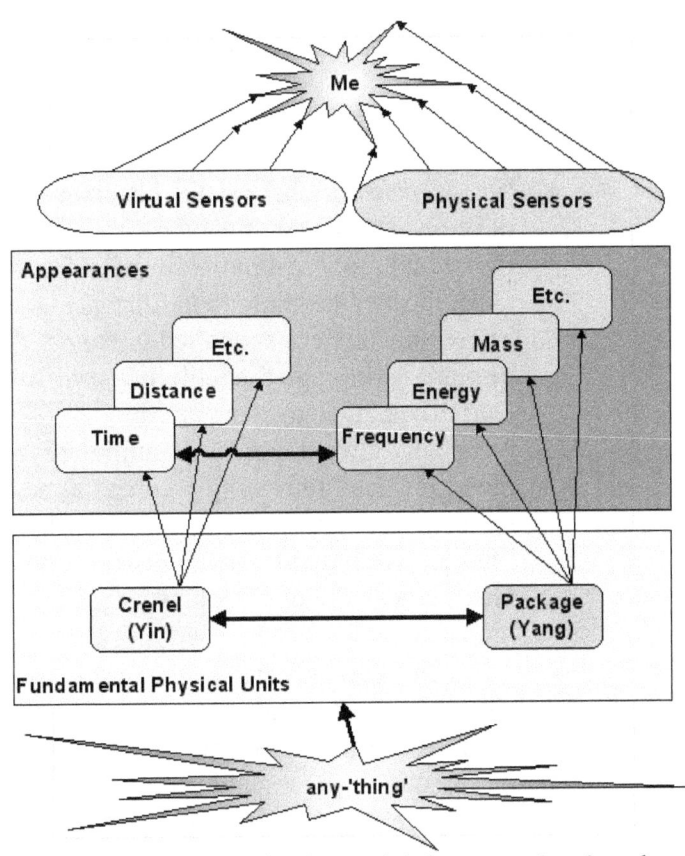

Figure 12.1: a basic model, between 'me' and any-'thing'....

One can follow the figure top-to-bottom, or the other way around in some more detail. Both routes will be described here.

The top-to-bottom sequence:

From the moment an individual human (indicated

as 'Me' in the above figure) is born, he gradually becomes aware of *two* types of sensors that – perhaps - did not exist prior to his physical conception as an observer (as a human being):

- 'virtual sensors' that –as the observer develops into an adult person– make the individual more and more aware of virtual appearances like 'time' and 'distance'. These appearances are called *virtual* because – although the individual will learn to quantify them - one cannot physically grasp them, nor manipulate them: for 'Me' they simply appear to be there. These sensors are related to the 'being conscious': if a person sleeps, he is much less aware of e.g. elapsing time.

- 'physical sensors' that –as the observer develops– made him more and more aware of *tangible* physical appearances like frequency, mass and energy. These sensors are tangible because one can associate them with organs that one can pinpoint on the human body: like eyes, ears, tongue, nose, skin.

  Even though there is no one-on-one relationship between physical sense organs and their properties (e.g. with the skin one can become aware of both frequency *and* energy), the tangible nature of these sensors is apparent.

These two types of sensors allow humans to become subjectively conscious of a surrounding world. After birth, one *gradually* becomes aware of distances and time, shapes, colours, mass, etc. Initially, the individuals 'science' is based on (and limited to) the awareness of the various signals that the sensors produce.

The two types of sensors that any human possesses connect him with two groups of *appearances*. Within each group, the appearances that one will actually perceive depend on the sensors that one is using. In fact, *all* possible appearances within a group are there, available and ready to be perceived or quantified or measured. This is regardless of the possibility that one's vision/perception may - in reality and in practice - be limited through the types and number of sensors that one has available (and actually pays attention to).

In Crenel Physics this leads to the hypothesis that – at bottom line – humans are monitoring no more than just *two* 'fundamental physical properties', named 'Crenels' and 'Packages'. The *combination* of these two properties gives awareness of the *where*-and-*what*-appearance of any-'thing'.

Any-'thing' can only appear to a person while it is containing both a 'Yin' *and* a 'Yang'. That is: an object needs to exist somewhere for some period of time (the Yin), *and* it needs to contain something (the Yang).

The 'bottom-to-top' sequence:

Here the reasoning starts with an object that is referred to as 'Any-'thing''.

In order for a person to perceive 'Any-'thing'', it must include both a 'world' around it (a 'Yin'), as well as contain some 'substance' (the 'Yang'). As reasoned in the previous chapter, 'Yin' and 'Yang' always -and by definition- come together. There *is* no 'Yin' without 'Yang'. And *both* are contained within 'Any-'thing'. (Note that there is yet another requirement to be met in order to allow perception. This requirement is discussed in chapter 14).

'Yin' is expressed in Crenels, 'Yang' in Packages. In Crenel Physics, Crenels and Packages are considered being the *fundamental physical properties*. They both come forth from an even more fundamental layer underneath, named: *any-'thing'*. This 'any-thing' is therefore the ultimate basis, and therefore representing the so called *natural unit* that comes forth from the here followed 'normalization process'. In Crenel Physics therefore there is only *one* natural unit... so far.

As discussed, Crenels and Packages are *reciprocal* units of measurement, and therefore related. In figure (12.1) this relationship is expressed through the forthcoming from their shared parent: *any- 'thing'* which is expressed in 'natural units'.

Both aforementioned fundamental physical properties can profile themselves through different *appearances*:
Crenels can appear to humans as 'time' or as 'distance'. That is so, because humans possess two separate and different virtual sense organs for 'Crenels'. And if an object indeed is observed, it appears to float around in some 3-dimensional spatial structure (set up in Crenels) as a function of 'time' (also set up in Crenels). Or: at some given time, the *any- 'thing'* is floating around somewhere.

Packages can appear to humans as 'frequency', or 'energy', or 'mass', etcetera. It solely and entirely depends on the sense organs (perhaps extended by physical instruments), what appearances one actually will become aware of in a particular case. The 'Packages' represent the content of the *any- 'thing'*.

The appearance 'Frequency' is a special appearance of the 'Package' because it sets the connection between Packages and Crenels. 'Frequency' is expressed in *cycles per Crenel*. In Metric Physics this connection is given by Planck's law ($E = h.\upsilon$).

In Crenel Physics this law has been fully integrated, which than resulted in the earlier conclusion that the *velocity of light* and *Planck's constant* are not to be seen as two separate universal physical constants: they actually are *one and the same* universal physical constant. Albeit that Planck's constant can only be applied when (and if) there is at least some non-zero substance around. This 'non-zero substance' has to be contained by 'Any-'thing''.

With this model and terminology in mind, the world of 'Any-'thing'' can be further explored. Obviously, such an exploration can best be done in the world of human *appearances*: this is the world one learned to perceive. The underlying world of Yin and Yang seems to be behind the horizon of human visualization. However, these more fundamental concepts should always be kept in mind.

13. Color intermezzo

The model as shown in Figure (12.1) in the previous chapter may –at first sight- be perceived as somewhat provocative: the entire human perception is classified as subjective, and thereby based on just two *fundamental* units of measurement (the Package and the Crenel) with only one *natural* unit underneath.

The question than is: where does this humanly perceived –beautifully diverse- world come from? How is this so? Why is this so?

That last question is outside the scope of this book, but the question *how* diversity manages to enter the subjective human arena (and even manages to present itself as if objective facts) can be addressed here.

According to Crenel Physics, humans cannot have an objective view on the physical world. This will be so, as long as the human perception is based on the types and qualities of the *human* sense organs (perhaps extended with scientific instruments) and on interpretations by the *human* brain. To a certain extent, humans unconsciously create their own perception of the world. This is rather dramatic because it implies that the human consciousness is -to a large extent- the product of the evolution of the human body and mind. And it also implies that other species –having different sense organs- would possibly perceive the same world in an entirely different manner.

To clarify how far this goes, and as an intermezzo to hard core Crenel Physics, this chapter discusses just one example of the impact of the human sense organs on the human perception. It deals with the human perception of *colors*.

Colors are just one of the results of the human interpretation of *Packages*. Like 'mass', 'energy' and 'frequency'. Color seeing illustrates that what (most) humans *see* is subjective, and cannot be objectively shared with other species (or with persons who are blind).

To analyze the 'content' that is represented by Packages, in previous chapters *Photons* were introduced into Crenel Physics. Photons carry light, and this light is an example of an actual and substantial 'thing' to humans. If Photons hit the eyes one can actually *see* light and its source. But the light 'as seen' is nothing but just one appearance of the fundamental property underneath: the Package. One will need healthy eyes to be able to perceive this particular appearance. And –as will be analyzed here– the actual light perception is multi-dimensional, colorful and indeed 'enlightening'. It will show that these appearances are indeed and entirely the product of the design of the *human* eye, and of human brain creativity.

From a physical viewpoint (alike in Metric Physics and Crenel Physics), 'light' is composed of individual *Photons*.

> Photons are - by definition – 'things' that travel at the 'speed of light'. Actually, for Photons there *is* no other velocity. Photons *are* light, so therefore it not a spectacular statement to declare that light has the speed of light… and thereby the speed of

Photons. No matter what an observer does, no matter where he is, no matter how fast he is travelling: if he measures the velocity of a Photon that is passing by, the outcome of his measurement is –by declaration- the same: he will find that the Photon has the 'speed of a Photon' relative to him. And this particular speed sets –in Crenel Physics- a standard. Or better: it *is* the standard. *And* it is dimensionless (see previous chapters).

In Metric Physics, this same speed of light is measured in *meters* per *second*. The complication here is, that a 'meter' is not always a 'meter'. The relative velocity between persons causes the meter to be relative (Lorentz contraction). And a 'second' is not always a 'second' (e.g. whether one is close to a gravitational mass, or not, has impact and makes the second to be relative as well). Thus, in Metric Physics the speed of light depends on circumstances… because it is measured in relative units of measurement. Only in deep empty space (vacuum) it is 'universally constant' and equals (approximately) 300.000.000 m/s.

Besides velocity, Photons have an additional property: each individual Photon also contains a certain amount of *energy* (or frequency, or mass) relative to the observer. In Metric Physics the range of energy (or frequency, or mass) that an individual Photon can contain, is referred to as the 'Electromagnetic spectrum'. And the amount of energy that a Photon contains – relative to the observer -

is expressed in various units of measurement, e.g. wavelength, frequency, EV (Electron Volt).

At bottom line, all these various units refer to one and the same underlying physical property: the contained energy (or frequency, or mass) relative to the observer. Therefore these properties can all be converted into *one single unit of measurement*. One could have picked any one of the aforementioned units, and declare it as *the* unit of measurement. The reason for using –in parallel- a series of *apparently* different units of measurement is based on history: pending the nature of experiments, a suitable unit of measurement was used. And pending the type of sensors one used, one quantified a different *appearance*. It took lifetimes of quarreling between physicists, before it finally and generally was agreed upon that indeed *all* these various appearances actually represent *one* – and no more than one – underlying fundamental physical property. In Crenel Physics this history is overseen, and one single and unique fundamental unit of measurement has been defined: 'Packages'.

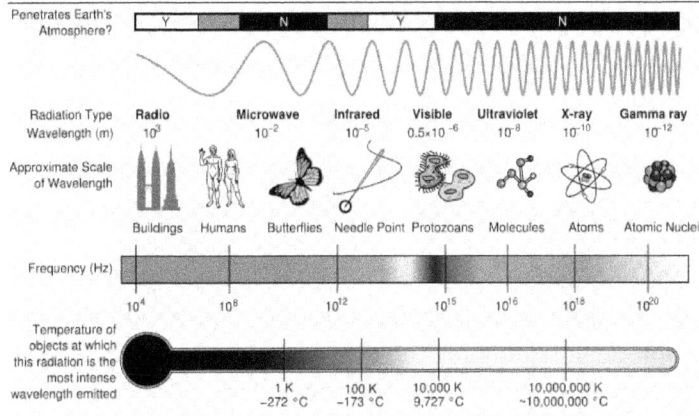

Figure 13.1: The Electromagnetic Spectrum: *Visible* Photons cover just a (small) range of the spectrum.(picture from Wikipedia).

Figure (13.1) shows the 'Electromagnetic spectrum', including various named regions. The entire range of the shown spectrum covers an impressive span: the Photons with *highest* energy contain 10^{16} times the amount of energy relative to the *lowest* energy Photons. This is a *huge* span. If the smallest Photon energy level would be represented by just one meter of rope, the largest photon of the shown span would be represented by 10^{16} meters of rope, which is about a light year of rope length. Or (as shown in the picture): the shortest wavelength of a Photon is about equal to the diameter of an atomic nuclei and the largest shown wavelength has the size of a building. And none of the shown energy levels could be considered rare: for all these shown levels huge amounts of Photons can be easily found (nowadays).

189

Just this single fact suggests that by exploring a 'Photon' one indeed is reviewing a phenomenon with broad relevance.

The figure also illustrates that - pending where an individual Photon resides on this spectrum - it may *appear* totally different. Gamma rays, radio waves, visible light… these are all potential *appearances* of individual Photons. For the current analyses the focus is on Photons that contain an amount of energy (or frequency, or mass), that fits somewhere into the – narrow- range that appears to humans as 'visible light'.

Figure (13.2) is an enlargement of this small section. It shows the associated visible *colors* that humans perceive, as function of the *energy* contained by a Photon relative to the observer.

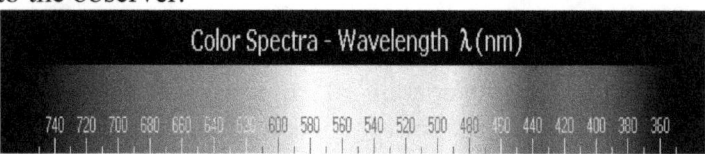

Figure 13.2: the colors of visible light.
(picture from Wikipedia)

It contains –assuming that it is printed perfectly- all colors of the visible spectrum.

The preliminary conclusion could be that therefore *all* possible colors are shown in figure (13.2). These colors indeed are equivalent to 'all colors of the rainbow'. And furthermore, figure (13.2) is a *one-dimensional* figure: there is only one (and no more than one) parameter or yardstick that determines the color of light that may be associated with a specific Photon. Here, this single parameter is indicated as the 'wavelength' of the Photon in nanometers. It represents the contained energy relative to the observer. And this is the physical property that has been named 'Packages' in Crenel Physics.

However, the above conclusion is indeed preliminary: the shown one-dimensional figure is *not* fully representing how human brains and connected sense organs actually deal with colors. One just needs to look at a piece of white paper to illustrate the point. If done so, one clearly sees *white* light coming from the paper... Thus, to humans, white light exists. In reality however, one sees something that does *not* exist as such. *There are no Photons that emit white light*. The color 'white' cannot be found in the spectrum of figure (13.2).

Being able to perceive the color 'white' is important too. Not only because it is just *one* additional color. If one mixes a little bit of color from the spectrum of figure (13.2) with a lot of this 'white', one perceives a whole new range of colors called 'pastel'.

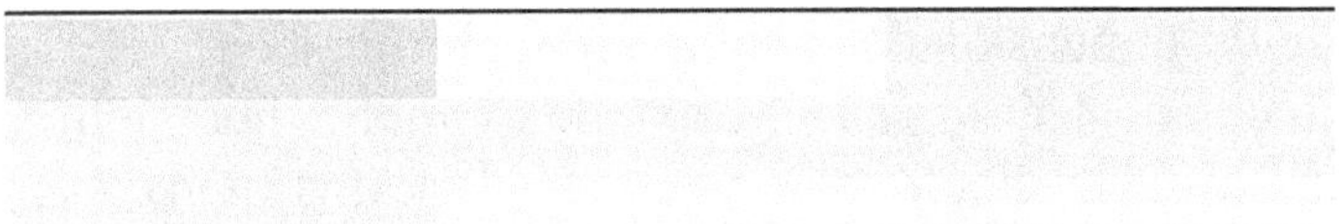

Figure 13.3: some pastel colors:
a little color from the spectrum mixed with a lot of 'white'.

As for the color 'white', one neither finds these pastel colors in the spectrum of figure (13.2). So what causes these *appearances* of colors?

Within human eyes (assuming one is not color blind) there are *three* different types of color receptors, each type exclusively sensitive to its own narrow color band (or: Photon energy band). The associated color bands are named 'red', 'green' and 'blue'.

The first and fundamental question thereby is: what exactly is a 'color'?

The physical answer is that *'color' does not exist…* 'Color' is nothing but an *appearance* presented to humans, by human brains. It is a subjective human perception of Photon energy levels within a narrow range.

And it is not just a 1-dimensional model that makes humans see colors. If that were the case, one would not be able to see white light or pastel colors.

How than is it nevertheless possible that humans can perceive 'white' light, and pastel colors as additional color ranges? Apparently, human brains manage to create a *multi*-dimensional perception model from a *one*-dimensional system. This capability enriches human life. But this also should raise a general alert: it is *subjective,* and exclusive to humans from the viewpoint of Crenel Physics.

Human eyes produce – per type of color receptor – an individual *color strength signal*. These signals add up to a mixture, in which each signal represents its own exclusively associated color band (red, green or blue). When all three color signals are of more or less equally balanced strength, humans perceive a *new* color: 'white'.

And, as 'red', 'green' and 'blue' are just *appearances* to humans, so is 'white' just an appearance.

First review the most basic possibilities of color-seeing. Assume that any of these three receptors can be either switched 'on', or switched 'off'. In reality each signal strength can vary in a more or less continuous range from value 0 (no signal) to value 1 (maximum signal). But here the extreme on/off possibilities are explored first. This 'binary' approach gives the most obvious five possibilities, as indicated in figure (13.4).

Red	Green	Blue	Result	
0	0	0	Black	
0	0	1	Blue	
0	1	0	Green	
1	0	0	Red	
1	1	1	White	

Figure 13.4: the five most obvious possibilities for color signals. 'White' is nothing but a balanced mixture of red, green and blue.

In the top line of the table, all three signals are switched off, and one would see no light. One perceives this as the color 'black'. 'Black' does not mean that no Photons hit the eyes. It only means that such Photons do not trigger any of the three types of color receptors in human eyes, because their energy level is outside the narrow energy range associated with visible light.

In the next three lines, only *one* type of color receptors is activated whereas the other two types are inactive. Thus, these lines give the respective *perceptions* of the - with the sensors associated - colors. And in the lowest line of the figure, all receptors are equally active: one perceives the 'new' color 'white'.

All three individual color *signal strengths* that are produced by human eyes can vary between 0 (no signal) and 1 (maximum signal strength). And when all three color signal strengths are increased -while maintaining their mutual ratios-, the perception of color remains unchanged. This is illustrated in Figure (13.5): a gradual and proportional increase of signal strengths results in various levels of intensity of 'white', referred to as a 'grey scale'.

Red	Green	Blue	
0	0	0	
0,2	0,2	0,2	
0,4	0,4	0,4	
0,6	0,6	0,6	
0,8	0,8	0,8	
1	1	1	

Figure 13.5: the color 'white' can come in various strengths.

There are three additional extreme –full signal strength-possibilities: the cases where *two* out of the three light signals are switched to their maximum, while the third signal is switched off. Thus, in effect the light of *two* colors is mixed. Together with the five on/off options already shown in figure (13.4) this results in a total of eight binary (or 'on/off') options, each associated with a specific color perception (including 'black' and 'white').

Red	Green	Blue	Result	
0	0	0	Black	
0	0	1	Blue	
0	1	0	Green	
0	1	1	Blue-Green	
1	0	0	Red	
1	0	1	Violet	
1	1	0	Yellow-Orange	
1	1	1	White	

Figure (13.6): eight possibilities for switching three color signals *on* or *off*.

These three additional on/off possibilities give the perception of respectively Blue-Green, Violet and Yellow-Orange, as shown. This completes the number of options that are based on the *binary* on/off approach. Note that the three newly added mix colors are indeed perceived as *new* colors, rather than as mix-colors (which they actually are).

The mathematics behind the above is: with three bits of information one can define 8 different statuses. Figure (13.6) thus encompasses – from a mathematical point of view - a 3-bit counter. To fully express the mixing possibilities of the aforementioned three colors, the table of figure (13.6) is too limited in its 1-dimensional layout.

With the analyses in mind, the entire range of potential color appearances can be represented in a *three* dimensional spatial model, e.g. a cubical box with rectangular coordinates X, Y and Z (or better: 'R', 'G' and 'B'), where each coordinate represents the strength of the light signal of the associated color. The strength of each signal could than vary between 0 (no signal) and 1 (maximum signal). Thus the point with coordinates 0,0,0 would represent 'black', and the point 1,1,1 would represent the brightest white light that eyes could see. Any individual spatial location within this cubicle would represent a unique combination of color and intensity.

The grey tones of figure (13.5) would then be found on the cubic's diagonal. And all pastel colors with a light 'hue' of color would be near this diagonal: the nearer to the diagonal, the lighter the hue.

If one takes the variable 'intensity' out of the equation (and just pick one fixed value) the entire world of possible colors would lie on the surface of a sphere centered around point 0,0,0. Note that only part of the surface of this sphere is used for the model: $1/8^{th}$ part of the entire sphere surface. This is because all three coordinates must have a positive value.

A random point within the cubicle does not 'feel' as a linear mixture of 'Red', 'Green' and 'Blue': human brains present a perception of distinguished *additional* colors. For example the color 'Yellow' feels as a separate color, clearly to be distinguished from 'Red' and 'Green'. While in fact 'Yellow' is a mixture of 'Red' and 'Green'. The color 'white' gives another perception. Although 'white' does not feel like a *color* or like a –balanced- mixture of colors, it actually *is*.

> 'Light' is apparently so important to humans, that a forth type of light receptors is built in the human eye: this type is operating independent from the other receptors. It is extra sensitive, and meant for dim conditions. These receptors give only one color perception: the equivalent of 'white' (on a grey-scale pending the intensity of the source).

In conclusion: although the appearance 'color' is in fact based on nothing but a *one*-dimensional physical system of energy levels of Photons, for modeling its *appearance* to humans a *3*-dimensional model (a cubicle box) is needed. And within this 3-dimensional system there is a mystery guest called 'white'.

Partially colorblind persons who are looking at an equally balanced color mixture (of their remaining visible colors) will still perceive this mixture as 'white'. Apparently the basic requirement for the appearance of 'white' is that the various color signal strengths that *are* received by the brains are in balance. Irrespective whether there are three different color signals available, or only two. This –again- illustrates that colors are a subjective perception: one person may see 'green' light while the other would judge it to be 'bleu'.

It may be coincidence that the human eye has *three* types of color receptors. The forthcoming color model thus –by this coincidence- required a 3-dimensional spatial representation to reflect all associated degrees of freedom. Such a model can be easily envisioned.

Hypothetical species that have *four* different color receptors (e.g. Red, Green, Blue *and* Ultra-Violet) would require a 4-dimensional spatial representation for their color model. Humans could not possibly envision such a model.

This chapter has focused on some specific type of appearances of the *Package* to humans. In the chapters to follow, the human multi-dimensional perception of 'time' and 'space' will be further discussed, thus with the focus on the *Crenel*. Where this chapter illustrated that 'red' is indeed nothing but a subjective appearance to humans and that one thus cannot argue that 'red' objectively *exists*, it was already concluded in a likewise manner, that one cannot really argue that a 3-dimensional spatial space *exists*. The latter was –likewise- presented as nothing but a subjective *appearance*.

It is appreciated that any human brain will have great difficulty in coming to peace with this outcome. Daily life and thinking loses its -originally thought- objective basis underneath. However, this fundamental basis is not gone: it is found one level deeper as indicated in the 'basic model' that was shown in chapter 12. This chapter is just one illustration of the point made.

The targeted *leanest possible* way to describe physics will –as was indicated in chapter 1- indeed require some major gymnastics of the human brain.

14. Detecting 'Any-'thing''.

This chapter and the upcoming ones deal with appearances that are associated with the Crenel. As per Figure 12.1, humans associate Crenels with 'time' and 'distance'. It was postulated that humans possess two associated *virtual* sense organs. Neither 'time' nor 'distance' can physically be grabbed. Therefore the sense organs that humans developed to become aware of these were called *virtual* (see chapter 5). These appearances do reside somewhere and in some form in the human consciousness, but are not generated through tangible signals.

To be able to detect any-'thing', two *objective* requirements –independent to human perceptions- were already identified:

- **Yin:** an object must exist for at least some non-zero period of time (which was earlier rephrased to: something must have moved around at least a little bit) within some spatial frame, and,
- **Yang:** it must contain a non-zero number of 'Packages'.

However, there is a *third* requirement.

This requirement is associated with the maximum possible velocity. In Crenel Physics this maximum velocity is a maximum conversion rate between Crenel appearances, and it has the numerical value of 1 (relative to any observer).

The concept behind this third requirement is, that at the moment an object is born it supposedly *cannot* possibly be detected *immediately, anywhere* and at *all times*. It cannot immediately be detected in the entire 'Yin'. Instead, at the moment an object is born it will start expanding its world of influence: its private 'Yin'. This expansion can only be achieved at the maximum possible velocity of 1 'Crenel' per 'Crenel', that is: the velocity of Photons (or: light). From a 3-dimensional perspective this expansion can best be envisioned through spherical coordinates: one must reside at (or reside within) a radius 'r_{max}' of an object's world, in order to be able to detect it. Here, 'r_{max}' – expressed in Crenels - appears simultaneously as a *distance* away from the object, as well as to the *age* of this object. Only through these two simultaneous appearances ('distance' *and* 'age/time') one can envision what is happening: the continuous increase of 'r_{max}'.

In Crenel Physics there is a one-on-one relationship between the numerical value for the 'age' of an object (expressed in 'Crenels') and the numerical value for the radius 'r_{max}' (also expressed in 'Crenels') of the spatial 'globe' from within this object can be detected. This one-on-one relationship is obvious because *both* are expressions of the same underlying fundamental physical property: the Crenel. If 'r_{max}' passes the observer at a certain moment in time, right at that exact same moment the observer will see the birth of the object, at a distance equal to the age of this object. The surface of the ever expanding globe represents the 'birth surface' at which an object is seen born. This 'birth surface' is represented in the two-dimensional figure below, in which this surface is represented as a 'birth line':

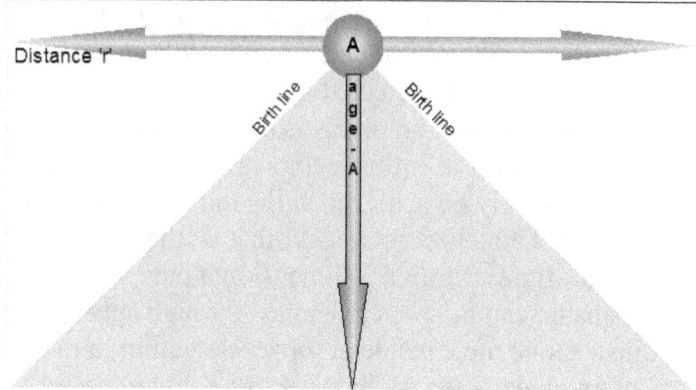

Figure 14.1: using spherical coordinates, the radius 'r' of the world of 'object A' expands one-on-one with its age: one Crenel per Crenel.

In daily life one may see a star being born, while looking at the nightly sky. Before one sees this happen, one had no means to detect that it existed. However, once one has seen it being born, one knows –based on its distance to the observer- how long its light was underway towards the observer, and what –consequently- the current age of the star must be (according to the observers timekeepers). Even from a subjective and local viewpoint the star *must* have existed *before* one had the possibility to detect this from a distance.

The above figure 14.1 graphically expresses the aforementioned 3rd requirement: with 'time' and 'distance' being the same physical property (both expressed in Crenels), the figure shows in a two dimensional manner that if one resides e.g. 10 Crenels away from 'object A', it will take 10 Crenels (after 'object A' is born) before one could notice it being born.

Therefore, it depends on the combination of the age of 'object A' and the distance 'r' away from 'object A', whether one can detect it or not.

Figure 14.2 (below) graphically illustrates the next step: what happens between *two* remote and presumably equal objects 'A' and 'B', that are born simultaneously at equal distance to the observer, and that presumably did not move from their respective relative spatial positions.

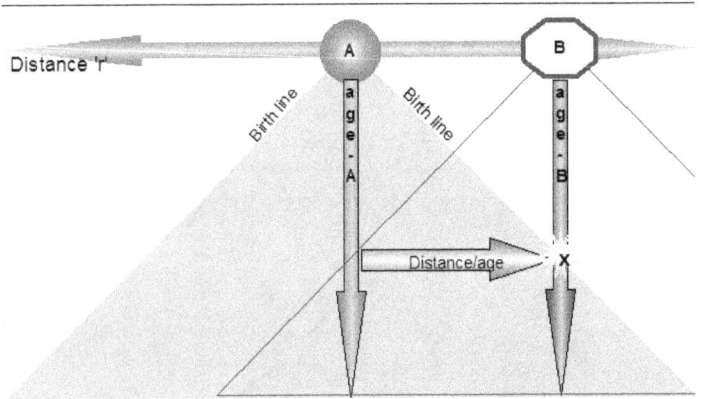

Figure 14.2: on distance/age marked 'x', the object 'B' sees the birth of object 'A'.

In this figure, at the age marked with 'x', the object 'B' sees the birth of 'A' at a distance which is – when expressed in Crenels - equal to its own age (also expressed in Crenels). Although at point 'x' the birth of 'A' is observed by 'B', at that moment 'A' is actually of the same age as 'B' (they were assumed to be born simultaneously). And because both A and B were presumed equal objects, the whole model in figure (14.2) is symmetrical. 'A' and 'B' can be exchanged vice versa.

By bringing 'A' and 'B' closer towards each other, the mutually observed age difference between 'A' and 'B' is reduced. Thus, the radius 'r' between the two objects represents both a relative age difference (the 'Time' appearance of the Crenel), *and* a distance (the 'Distance' appearance of the Crenel).

Due to the finite maximum light velocity (of value 1), interactions like the gravitational force require *propagation* between 'A' and 'B', and vice versa, before they become effective.

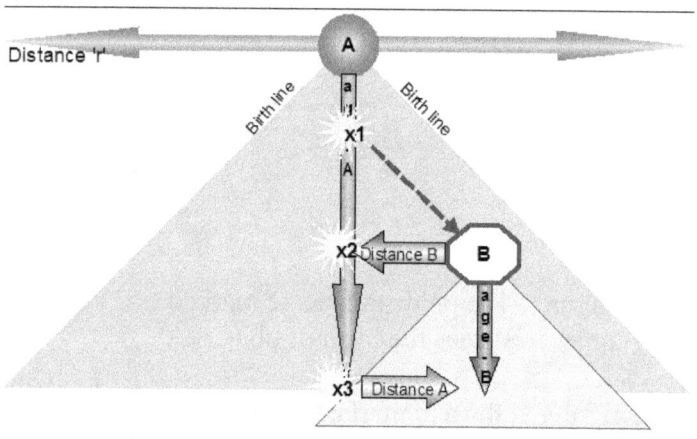

Figure 14.3: object 'B' being born within the 'Yin'-world of object 'A'.

Figure (14.3) illustrates some key consequences in another –third- scenario. Here, *after* 'A' was born and started to expand its world, 'B' is shown to be born *within* the world of 'A'. Object 'B' therefore experiences the presence of 'A' *immediately* upon its own birth. One consequence in this scenario is that –seen from a remote viewpoint- 'A' apparently was already exercising a gravitational force on 'B' at an age *before* 'B' even existed... That is: starting from 'A's own age (the moment) marked 'x1' onwards. This happened *without* experiencing the associated reaction force...

To clarify the point: assume a particle being born on Earth. It will immediately be subject to the Sun's gravitational pull. However, it will take about 8 minutes before the Sun becomes aware of the fact that the newborn particle started pulling 8 minutes ago...

Apparently and at first sight, the law of 'action and reaction' is violated here, at least for some period of time.

While 'A' is exercising a gravitational action force on 'B' from moment 'x1' onwards, figure (14.3) makes clear that the associated reaction force starts working only from moment 'x3' onwards. Because it takes until moment 'x3', for 'A' to notice that 'B' is born.

The scenario that is expressed by figure (14.3) would – by an external observer – be described as follows:
- in the age (timeframe) from 'x1' until 'x2', object 'A' was exercising a gravitational force on a thing that did not even exist,
- in the age (timeframe) from 'x2' until 'x3', object 'A' was still exercising this force on an existing 'B', while the presence of 'B' still did not yet

become detectable by 'A'. But during this timeframe, 'B' would de facto exist from a remote point of view.

- From moment 'x3' onwards, 'object A' will see the birth of 'B', and simultaneously start experiencing the reaction force related to the action force that it applied from moment 'x1' onwards. This reaction force will appear to come from 'B'.

- Should – in the further future beyond point 'x3'– object 'B' seize to exist (this is not shown in the figure), the reaction force on 'A' would continue for a while.

Therefore, in this sequence of events the *apparent* violation of the law of action and reaction – when integrated over time - will ultimately not materialize.Figure (14.3) shows, how coordinates of type 'Distance' must be capable to carry and contain both a propagating force and a propagating reaction force for some period of time.

> *Coordinates of the type 'Yin' (time or distance, expressed in Crenel) can 'contain' propagating gravitational forces that can materialize in the future (or: materialize somewhere else).*

In this example there is only one 'Distance' from the perspective of the observer: the line that connects 'A' to 'B'. This one dimensional connection line contained - for some period of time - an action force and a reaction force, while these were propagating.

The here described mechanism that an object can *currently* influence its own future in a predictable way *without any possibility* of 'knowing' this *as it is happening*, is something that does not fit into the human daily logics. Philosophy and physics touch each other here. The next chapter further investigates this mechanism, using scenarios with moving objects.

Prior to that, there is reason for further analyses of the here described static model.

In earlier chapters the concept of a 'Photon' was introduced as: a container for 'Packages' traveling at the speed of 'Photons', which speed is also referred to as the speed of 'light'. In modern physics, this 'speed of light' has a special load: it seems something hard to envision in that this is a *maximum* speed. It was rightfully introduced as something special. In Crenel Physics the implications have been shifted: of course Photons have the speed of Photons… and therefore this speed will be used as reference to all speeds.

In this chapter it is concluded, that Crenels *also* can contain something else besides Photons, namely propagating (gravitational) forces. This is because these forces are propagating at finite speed, presumably the speed of 'light'. The possibility to contain a propagating force is a remarkable feature. One question is, whether this is a *new* feature, or whether this is a logical consequence of what has already been defined in the Crenel Physics toolbox. Note that the objective is to *not* introduce new features into Crenel Physics, unless they indeed bring something new.

In Crenel Physics, the unit of measurement for 'force' is: P/C. Therefore, using figure (14.3), the aforementioned feature of Crenels can be rephrased as follows:

> *Crenels can contain a certain number of P/C's (read: 'forces') for some Crenels (read: period of time).*

> *If such a contained force is between two objects 'A' and 'B', the presumed duration of this containment is equal to the number of Crenels between 'A' and 'B'.*

The above statements gain credibility if the earlier conclusion - that Crenels are reciprocal to Packages and therefore related to Packages - is taken into consideration. Perhaps, the concept of temporarily containment of forces also forms the basis for 'dark matter', and an effect called the 'Tunnel effect':

'Dark matter':
Astronomers struggle with gravity forces that apparently are around, while the associated matter cannot be found. The Crenel Physics explanation than would be, that matter once indeed 'existed' in our universe and initiated propagating gravity forces, but that this matter meanwhile shifted into other appearance(s) that are not detectable. The still propagating gravity forces would be logical leftovers.

'Tunnel effect':
According to this effect, particles can take barriers that they should not be able to take. Or differently formulated: particles can disappear from one spatial

location, and re-appear in another spatial location where they – according to the anticipated applicable forces en route – could not arrive. In between – while taking a seemingly impossible barrier – they are nowhere around. According to the above model, the applicable forces -that one would expect- would be temporarily contained in Crenels, thereby allowing a particle to pass the seemingly impossible barrier.

When forces can be temporarily contained by Crenels, they thereby temporarily may disappear from the apparent world as this is seen by some observer, to reappear after some Crenels (that is: somewhere else, at some other time).

The question than is whether 'B' (see Figure 14.3) indeed and truly was 'away', prior to its appearance to the observer. Was it indeed *non*-existing prior to the moment it was seen being born? It seems logical to presume that it actually existed in some appearances *all the time* (and therefore *anywhere*), so that prior to its birth *as a Package containing* object in the apparent world, it was existing in some other form. In Crenel Physics this mechanism could require a transformation from Packages into Crenels (there is nothing else…). Because Packages and Crenels are reciprocal to each other, there *is* a relationship, and such a transformation cannot be excluded. Such a thesis would also support the universal principle of containment of mass and energy. One could –as an alternative- accept that from a universal perspective other encompassing 'worlds' exist, 'worlds' that are not apparent to a particular observer. Given the perspective of Crenel Physics, and referring e.g. to the model discussed in the previous chapter, the acceptance of such non-apparent worlds is an option.

Relate all the above to the 'potential energy' of a stone that one holds one meter above the ground. This 'potential energy' *is* contained somewhere, but it will only reveal its existence in the observer's world when it is dropped.

15. The Moon, the Earth, and the Notohp.

This chapter further illustrates the impact of the qualification '*apparent* world', being a human interpretation of the fundamental reality underneath. Evaluations based on *appearances*, on 'the looks of what one observes', give more insight if the underlying fundamentals are better understood.

As an example, the orbits of the Moon and the Earth around their central point of mass will be analyzed, thereby using Metric units of measurement. As discussed, these Metric units themselves *are* appearances... of the underlying fundamentals: the Package and the Crenel.

To simplify the upcoming analyses, it will be assumed that
>   1. *The Earth and the Moon are of equal mass, and*
>   2. *Very small in size.*
Thus, as point shaped objects they both orbit with equal orbit radius around a shared central point of mass that lies in the middle between them. These two simplifications have no conceptual impact on the analyses.

It will become clear that - when expressed in Metric units of measurement- there are various and different *appearances* of this orbiting model. These are relative to (and pending on) the observer's perspective, and therefore these appearances are *not* absolute. The 'absolute truth' lies *under* the various appearances that Metric Physics presents.

Despite this variety in appearances, at the bottom line there obviously is only *one* orbiting system here. In Crenel Physics –being based on nothing but fundamental and 'absolute units'- one therefore can rightfully expect that there is only one –and no more than one- way to describe this model. It is this particular and unique description that is referred to as being the 'fundamental truth'. This truth must be based on *fundamental* units of measurement, thereby encompassing (or: inherently including) the effects of relativity.

As result of the upcoming reasoning, a relationship between several Metric yardsticks will be quantified in the form of a conversion factor for the various apparent – and thereby relative- perspectives. In Metric Physics this conversion indeed is part of the 'theory of relativity' (the relativity of Metric yardsticks).

A *remote observer* is now introduced into the aforementioned Earth/Moon orbit model that is under consideration here. He resides on the North Polestar, which is located at a point very far above the shared central point of mass, and which also is located perpendicular to the plane in which the Earth and Moon orbit. Figure (15.1) shows, what this remote observer sees from his perspective.

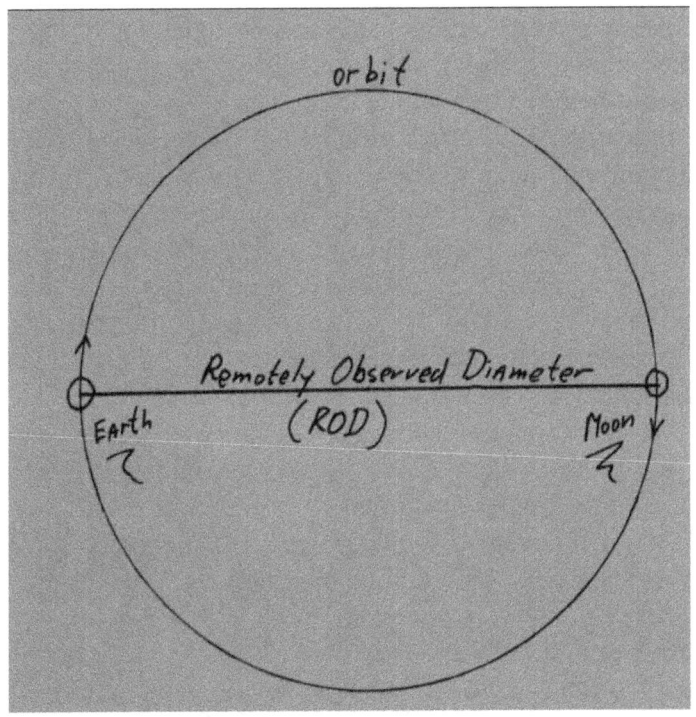

Figure 15.1: Earth and Moon in their shared orbit, as seen by the remote observer on the North Polestar.

This remote observer sees the Earth and Moon orbit along the same shared orbit path, which has a diameter indicated as the 'Remotely Observed Diameter' or 'ROD'. Thereby, the Earth and the Moon are –at any time- at opposite points in their shared orbit.

One advantage of this remote observation point is, that - at any moment in time- it will be exactly as far away from the Earth, as it will be from the Moon. Therefore, if there are events on either the Earth or on the Moon, the remote observer will see these happen equally retarded. The 'retardation time' is the time that Photons (=light) require to travel from either object -Earth or Moon- to the remote observer. Thus, the retardation of the observations is a constant factor to this remote observer. The sequence of events 'as seen' will be used *as a convenient reference* for the 'objective' sequence of the happenings... a while ago... on either Earth or Moon.

The observations of this remote observer are *selected* as reference (or: base case) for further evaluating the dynamics from other perspectives. This choice is indeed a free choice: other points of reference could have been selected as well, giving another base case for events on Earth or Moon being *simultaneous*, or taking place at some sequence within some time interval.

The length of the 'Remotely Observed Diameter' (or: ROD) of the shared orbit –as seen by the remote observer- is an important physical model constant. It dictates for example, how fast communication between the orbiting objects (Earth and Moon) can be from the remote perspective.

To envision such communication, assume that a person on Earth switches on a light. According to the remote observer it will take a while before this can be noted by an astronaut on the Moon, because the light beam (Photons) requires some travel time to cover the distance 'ROD' between Earth and Moon.

To acknowledge arrival of the light beam, assume that – as soon as the astronaut sees the earthly light burning- he will signal this by switching on a light on the Moon. This simple procedure describes the most basic form of interactive communication (or: interaction), performed in the fastest possible way. By switching on his own light, the astronaut gives a very basic message: 'yes, I now see that the light on Earth has been switched on'.

The *remote observer* sees both actions equally retarded: first, at some moment in time, he will see the light on the Earth being switched on. Using his Metric units, after a delay of about 1.3 seconds thereafter he sees the light on the Moon being switched on, indicating the arrival of the Earthly light. The distance between the Earth and the Moon (the ROD) is approximately 400.000.000 meters, and light travels at a –constant- velocity of approximately 300.000.000 meters/second. This explains the aforementioned approximate 1.3 seconds delay time.

As was discussed in previous chapters, these 1.3 *seconds* also is a measure for the *distance* between Earth and Moon. As the aforementioned 400.000.000 meters is a measure for this same distance. The 1.3 seconds is a distance measurement which is based on the watch of the remote observer on the North polestar, and on light velocity being a known natural constant which is equal to all.

The usage of the 1.3 seconds as a measure for distance is based on the presumption in Metric Physics, that the velocity of the orbit crossing light indeed is a universal constant. The latter is a confirmed fact in Metric Physics, for light crossing empty spaces (vacuum). Note that in Crenel Physics this same velocity of light is set as *the* reference for measurements. Therefore the aforementioned *confirmed presumption* in Metric Physics is promoted to an *inherent truth* in Crenel Physics: a truth by implication. Mainstream physicists will have no problem with this approach.

Standing on the Earth gives another frame of reference (or: perspective) to the model. From this position one would see a local event on Earth immediately… as it happens. However, an event on the Moon –like switching on a light- would be seen retarded. This is –again- because Photons (or: light) leaving the Moon require about 1.3 seconds to reach the Earth.

When –from this Earthly position- a person is looking at the Moon, he does however –according to the remote observer's frame of reference- *not* look into the direction at which the Moon is *at that moment.* The remote observer will analyse this scenario as follows: while the Moon is orbiting, the Earthly observer is reckoned to look at a *retarded* position of the Moon, into a spatial direction where the Moon *was*… about 1.3 seconds ago. During those 1.3 seconds the Moon moved approximately 750 meters further along its orbit path.

This difference of 750 meters, between where the Earthly observer is reckoned to *see* the Moon, and what actually can be reckoned that actually *is* the Moon position (all according to the remote frame of reference that was selected as *the* reference), is small relative to the much larger (approximate) 400,000,000 meters of distance between the Earth and the Moon. But the mechanism that causes this small but apparent difference becomes of importance in cases where orbit velocities are higher. Because this slow motion model of the Earth/Moon system is relatively easy to envision, it is selected here for further analyses, prior to a discussion of the higher velocity cases (see the next chapter).

To further explore the above observation, assume now that an astronaut has installed a small mirror on the Moon. The objective is, to hit this mirror with a Photon that is fired with a laser gun from Earth. In this setting it is clear that - should the gunman on Earth aim his laser gun to the position where he actually *sees* the mirror (and Moon) - he is pointing 750 meters away from where the target actually *is*: the mirror currently is *not* at the - metric- coordinates where it is seen.

Knowing the orbit path and the orbit velocity of the Moon, the gunman could therefore decide to aim the laser gun 750 meters away from the target, in the direction of the Moon's orbit path. That is the position where he reckons that the mirror physically *is*. So now he fires a Photon. Despite this correction, the target will still be missed because it takes the fired Photon about 1.3 seconds to reach the mirror. And during those 1.3 seconds the Moon -and mirror- moved another 750 meters along their orbit.

Therefore, the gunman must neither aim at the coordinates where he *sees* the mirror, nor at the coordinates where he reckons the mirror physically *is*. Instead, he must aim at the point where the mirror is reckoned to be in the future, by the time the fired Photon reaches the Moon's orbit. The gunman therefore must aim about 1500 meters away from where he actually sees the mirror. The scenario is indicated in the next figure:

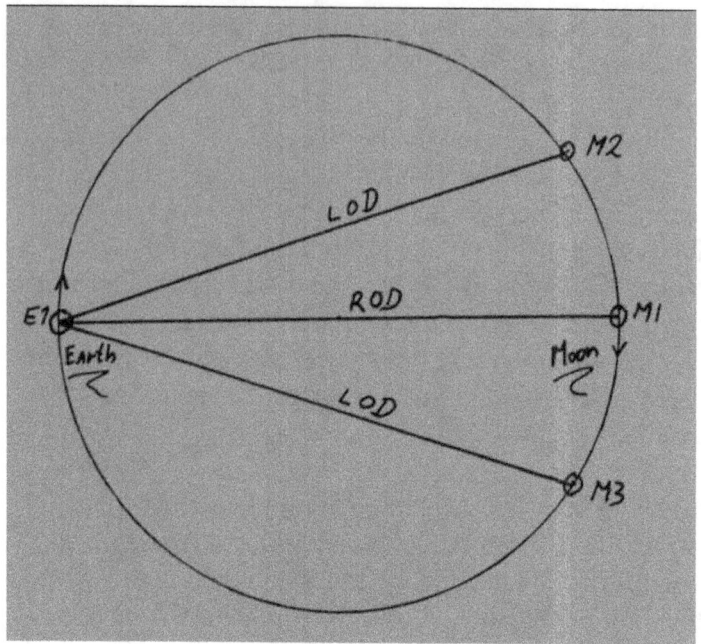

Figure 15.2: three respective Moon positions.

In figure (15.2) the three discussed Moon positions are shown: M1, M2 and M3, each about 750 meters apart:

- M1 is the actual Moon position according to the remote observer,
- M2 is the retarded position where the gunman will actually see the Moon,
- M3 is the future position to aim at.

Figure (15.2) shows these three Moon positions from the perspective of the remote observer on the North Polestar. This is, because the remote observer's observations were selected as the base case for further analyses. From this position one can –as done- easily envision the challenge of hitting the mirror, and the thereby given considerations.

The lengths of the two orbit sections between M1 and M2, and between M1 and M3 respectively, both were *estimated* to be about 750 meters. The *exact* lengths are defined as follows: it is the length of the orbit path that the Moon would travel in its orbit, while a Photon *crosses* the orbit from the Earth to the Moon.

For reasons of symmetry the indicated section lengths M1M2 and M1M3 are *exactly* equal to each other. Note that in figure (15.2) –and in the following figures- these two orbit sections of 750 meters have been over-scaled relative to the diameter of the orbit: the indicated 'ROD' (Remotely Observed Diameter) is about 400,000,000 meters. This over-scaling is done for graphical clarity only.

Thus, in this model there now is:

- The 'gunman' with a laser gun on Earth,
- The 'astronaut' with mirror on the Moon, and
- The 'remote observer' on the North Polestar.

Figure (15.3) shows the discussed *Photon paths* from the perspective of the remote observer. He sees -at a certain reference moment in time- the Earth at the position E1, and the Moon at the exact opposite position M1.

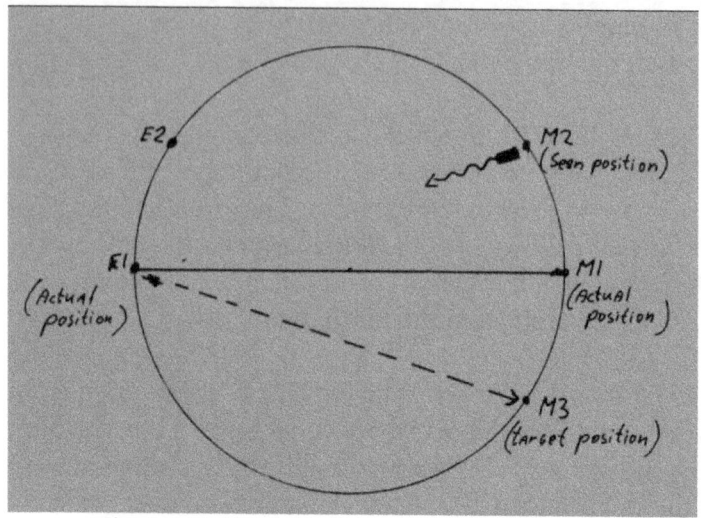

Figure 15.3: 'seen', 'actual' and 'target' positions of the Moon, shown within the remote frame of reference.

While the gunman stands on the Earth (at the reference position E1), the remote observer will reckon that this gunman will actually *see* the Moon at position M2. This is because he reckons that the light emitted by the Moon - while at this position M2- reaches the Earth at the moment the Earth arrives at E1. According to his remote observations -at that moment- the Moon's *actual* position is further along the orbit: at M1. And in order to hit the mirror with his laser gun, the gunman must aim at target position M3. By the time the Photon reaches M3 (1.3 seconds after firing), the Moon also has physically reached this position.

There are two points to notice here:

1. **Orbit diameters 'ROD' and 'LOD'.**

Besides the already mentioned 'ROD', the earlier figure (15.2) also shows a 'Locally Observed Diameter' of the orbit, which is the distance between E1 and M2. It is indicated as the 'LOD'.

The remote observer –using his frame of reference- rightfully reckons that this 'LOD' is indeed the orbit diameter as it will be seen by the gunman, because the gunman –when in position E1- will actually see the Moon at the retarded location M2 (a location in the remote frame of reference).

This remotely reckoned 'LOD' is *shorter* than the remotely observed orbit diameter (indicated as 'ROD'). The difference between ROD and LOD is –at the bottom line- caused by the Photons requiring some time to cross the orbit, or: by their velocity being *finite*.

The distance from E1 to M3 (the 'target' position) is exactly equal to the 'LOD' (from E1 to M2). This is because the distance between M1 and M2 is equal to the distance between M1 and M3, resulting in geometrical symmetry. The 'LOD', being shorter relative to the 'ROD', is therefore also a *constant* in this model, in which the orbits are presumed to be circles.

Thus:

*One now has to deal with **two** different orbit diameters: the orbit diameter 'ROD' that one sees remotely from the North Polestar, and the orbit diameter 'LOD' that –from this remote position- is reckoned to be seen from Earth.*

The latter is shorter.

2. **Direction.**

The *remote observer* rightfully reckons that the fired Photon hits the mirror when the Moon arrives at point M3. While the Photon travels from E1 to M3, the *Moon* travels from M1 to M3 and the *Earth* travels from E1 to E2, as indicated in figure (15.3).

When the Earth reaches E2 –that is the moment of Photon impact on the Moon- the astronaut on the Moon *sees* the Earth in its 1.3 seconds retarded position: point E1 in figure (15.3). That was the Earth's reckoned position at the moment the Photon was fired.

Thus, the astronaut on the Moon will conclude that the arriving Photon followed a *straight* path through space: the Photon's incoming direction matches the direction towards its source: the direction at which the laser gun is *seen* at the moment of impact at the Moon. Therefore, to the astronaut the Photon appears to have followed the path from E1 to M3, per Figure (15.3), of which the length is also equal to the 'LOD'.

All the above locations are shown in the remote frame of reference, which was selected as the base case. Given the above considerations, the gunman on Earth can –from his position- reconstruct the shown figures. This requires awareness of the fact that the Earth on which he stands is actually orbiting around a shared central point of mass, with the Moon at the opposite point of the orbit. His general reasoning while shooting at a target is:

If one aims a laser gun at a target where it is **seen***, and the outcome is that one nevertheless misses the target, the conclusion is that this target must be moving into a tangential direction relative to the Photon's path (as would be the case when the target is orbiting).*

One can *not* reverse this general reasoning: 'if a seen target *is* hit by aiming into the seen direction, this only *could* imply that the target is not moving in a tangential direction'. If the target is hit, this *could* imply that the target completed exactly one (or more) full orbits while the fired Photon travels from the laser gun to the –initially- seen target position.

Therefore, to absolutely ensure that a target is *not* moving in a tangential direction, from a strictly mathematical viewpoint one would have to shoot a virtual infinite number of times at a virtual infinite number of different time intervals, *and* hit the target in all cases. This strategy is related to Shannon's theorem. However, by using physical rules on top of that, in order to get such a positive confirmation, the number of shots can be limited to only two shots at very short time interval. Within such a very short time interval one full orbit (or more orbits) could not possibly be completed because the orbit velocity of the target is physically limited: it cannot exceed the velocity of light.

Both the gunman and the remote observer come to a joint conclusion: when the Photon is fired towards the Moon, it only needs to *physically* travel a distance LOD, while – according to the base case frame of reference- the longer ROD is the actual distance between both orbiting objects.

With the LOD being shorter, it could now be –wrongly-postulated that consequently the gunman can communicate to and fro slightly *faster* with the astronaut on the Moon, than the –longer- remotely observed distance (ROD) between Moon and Earth would suggest.

To further investigate this –wrong- postulation, assume that the astronaut on the Moon wants to reflect the incoming Photon straight back to the gunman on Earth. Again, this can be modelled from the remote perspective, which serves as the base case here. Because the Earth is an orbiting target, the astronaut should get the instruction to aim his mirror towards the reckoned *actual* Earth position E2 (he does not *see* the Earth there yet... but he can reckon it is at E2 using gravitational orbit properties).

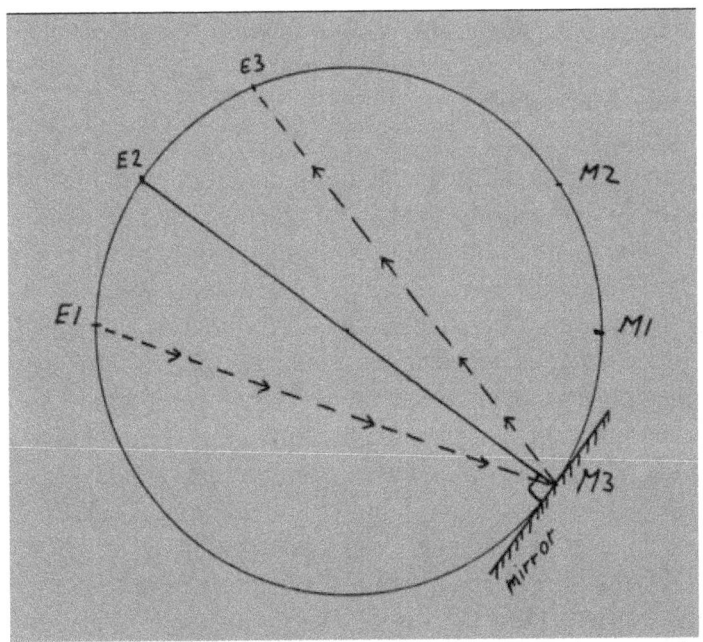

Figure 15.4: Reflecting the Photon back to Earth.

As indicated in Figure (15.4) and as discussed before, the astronaut indeed sees the Photon that was fired originate from point E1. Thus, when he aims his mirror to the reckoned *actual* position E2, the Photon will be reflected into the direction of E3. This is because -when a Photon hits the mirror- the incoming angle will equal the outgoing angle.

From the perspective of the remote observer based on the North Polestar, the exact *same* 'mirror aiming instruction' to the astronaut would be given. Although the remote observer sees a larger distance between Earth and Moon, the *direction* towards this 'actual position' is equal to, and shared with the astronaut's observation: it is the exact opposite point relative to the current position in the orbit. This physical direction towards the *opposite point* in the orbit is therefore shared, regardless the size of the observed object distance.

From both perspectives, the aiming instruction to the astronaut indeed must and should be the same: there is only *one* mirror, and therefore there can only be *one* direction to aim this mirror into. And -therefore- this instruction *must* be unambiguous. And it is: *aim at the reckoned actual physical location according to your own frame of reference*. This reckoned actual position –in all cases- is the exact opposite position in the orbit, relative to the position at which the mirror currently resides. It thereby is indeed not relevant that the remote observer sees a larger distance between both orbiting objects (the ROD) than the astronaut will see from his perspective (the shorter LOD): all aiming instructions indeed lead to *the same direction*, and therefore this direction is *absolute* (in this particular model).

Photons contain mass and therefore produce impulse on impact. The incoming path combined with the reflected path (with the mirror aimed towards position E2) will result in a net impulse force that is precisely directed away from position E2, which is the Earth's *actual* position at Photon impact on the Moon. It is an important insight that the aforementioned *absolute* direction coincides with the direction of the impulse force. The dynamics of the orbiting process (the apparent radius and the associated orbit velocity) may differ in various scenarios for this model, but these are irrelevant for determining the *absolute* direction that the impulse force will have. Or: the direction at which the Moon is actually *seen* may differ (and will differ) in different possible orbit scenarios: this *seen* direction is however not of relevance for the direction that the impulse force will have. The net impulse force will -in all scenarios- come from the exact opposite point of the orbit, albeit that the distance to this opposite point will appear different from various perspectives (ROD, or the shorter LOD, or perhaps other distances as seen from other possible perspectives).

In all model scenarios, the Photon is coming in at some – here small- angle on the mirror's surface (relative to a perpendicular direction). Consequently, the Photon's impulse force on the mirror is slightly lower, relative to a scenario in which the Photon is reflected straight back into the direction where it came from. In the current model the incoming angle is very small due to the relatively low orbit velocity. But in higher velocity cases (that will be explored later) this 'impact angle effect' on the impulse force will become more relevant: the higher the orbit velocities, the larger the Photon impact angle, and the smaller the impulse force on the mirror will be. The extreme case being that the Earth would travel 90^0 of its orbit path while the Photon travels from the Earth to the Moon: in that extreme case no mirror would be required because the Photon would come in parallel to the mirror's surface… But, in the absence of a mirror, how could the Photon than re-arrive back on Earth? These higher velocity scenarios –and this related question- will be addressed in the next chapter.

To enhance the current slow motion experiment, after the Photon is fired from the Earth, the laser gun on Earth is now replaced by a second mirror. When this second mirror -on Earth- is aimed towards the reckoned *actual*

Moon position, the Photon will now forever bounce back and forth between the Earth and the Moon. This leads to a continued repelling impulse force that comes in bounces.

> Note: if a *stream* of Photons were involved in this experiment, this would result in a (more or less) continuous repelling force between the Earth and the Moon, tending to separate them.

The strength of this impulse force:

1. Is proportional to the Photon's mass (or frequency, or energy containment). The higher the Photon's mass, the higher the repelling impulse force.

2. Is reciprocally proportional to the *distance* between the Moon and the Earth.

    The shorter this distance, the higher the bouncing frequency and the stronger the repelling impulse force.

    The remote observer can calculate a bouncing frequency based on the remotely observed orbit diameter ROD and on the -universally constant-speed of Photons. The ROD *is* the observed distance between both orbiting objects. And this ROD sets the maximum pace at which to and fro communication can take place between the Earth and Moon, based on the remote observer's clock.

    In doing so, he would however ignore the

dynamics of the orbiting system: he could rightfully reckon that the crossing Photon physically will *not* bounce along the path of this remotely observed diameter 'ROD', but instead it will follow the shorter path from E1 to M3, and from M3 towards E3, as per figure (15.3). Nevertheless, the remote observer *sees* this larger 'remotely observed diameter' between the Earth and the Moon, and therefore this distance –and associated bouncing frequency- has relevance in terms of how fast interactive communication between both objects can be, from his perspective (and measured on *his* clock).

He can also perform a *second* calculation for the bouncing frequency, this time based on the shorter locally observed orbit diameter LOD (per Figure 15.2). This is the distance between Earth and Moon, as it will appear to the local observers (on Earth and Moon). This second calculation would obviously result in a *higher* bouncing frequency to be applicable to the local observers (the astronaut and the gunman). This second calculation is –again- based on the assumption that the velocity of the crossing Photon is set as a *universal* standard, and therefore equal to all. And it is also based on the remote observer's clock.

To explore the difference between the two calculated results, the remote observer now asks the astronaut to send him an Email each time the Photon hits the local mirror on the Moon. Despite the higher bouncing frequency –found in the second calculation- these Emails would nevertheless arrive at the pace that corresponds to the aforementioned *first* calculation. The – perhaps- surprising element in this may be that the remote observer knows that the Photon physically does *not* follow the 'ROD' which was used in the first calculation: as mentioned before, the remote observer rightfully reckons that the travelling Photon physically follows the shorter LOD per figure (15.2).

One explanation -*why* the first calculation nevertheless dictates the pace at which Emails arrive- is that a 'full orbit' is a 'full orbit', from *any* perspective. After the Earth has completed a full orbit, it will be back at its original position E1 and all original coordinates in the model are restored. Thereby it is not relevant if the distance between objects is seen shorter (the local perspective) or longer (the remote perspective). The dynamics of a full orbit rotation as such is not relevant: a 'full orbit' is indeed a 'full orbit', from all perspectives. Relevant is, that after one full cycle the exact original coordinates have been restored from *any* perspective, *and* that thereby a certain –equal to all- number of Emails have been produced and sent. To clarify this point numerically (the exact numbers are not relevant here): the Moon orbits the Earth every 28 days, during which every 2.6 seconds an Email would be sent to the North Polestar. This results in 930.461 Emails per full orbit. It is now regardless of *where* (in which frame of reference) a recipient of these Emails resides: in all cases this recipient will receive 930.461 Emails per completed orbit.

Consequently, because these Emails:

a. *Do* appear to arrive at different paces and,
b. The velocity of Photons (as used in both calculations) is presumed equal for all,
c. While nevertheless after each 930.461 Emails a full orbit has been completed…

…there must be a difference in the paces at which the *clocks* –used to measure the bouncing frequencies- appear to be running. Apparently, each frame of reference has an associated yardstick for *time* measurements (seconds).

To further explore this apparent difference in time measurement numerically, assume that -based on the 'ROD' associated with the remote frame of reference- a Photon's round trip between the Earth and the Moon would last exactly $(2 \times 1.3 =)$ 2.6 seconds. From the remote perspective a shorter distance LOD between objects is reckoned to appear to the astronaut on the Moon, and thus – at universally equal Photon speed for all- the second calculation results in a slightly *shorter* round trip time of e.g. 2.599 seconds. The latter number is arbitrarily chosen here: its exact and correct value is not relevant now (as long as it is slightly less than 2.6 seconds). Thus, the astronaut will send these Emails every 2.599 seconds, according to his own watch. However, at the remote position the Emails do not come in every 2.599 seconds, but every 2.6 seconds instead. The smaller timeframe of 2.599 seconds on the astronauts watch corresponds to 2.6 seconds on the remote watch.

Therefore, relative to the remote observer, the astronaut's watch on the Moon (and likewise the gunner's watch on Earth) appears to run at slower pace.

3. (the repelling force) is also reciprocally proportional to the clock slowdown per the above point 2:

    Now that the remote observer is understanding that the local watches on both the Earth as well as the Moon appear to run at slower pace relative to his own watch, he will also reckon that the locally observed frequency (and thereby: mass containment) of the orbit crossing Photon *itself* will be judged differently: when the remote observer sees a Photon frequency of e.g. 2.6×10^{16} Hz, he can now rightfully reckon that on Earth or Moon this exact same frequency would be found proportionally *higher*, in this example $(2.6/2.559) \times (2.6 \times 10^{16})$Hz $= 2.601 \times 10^{16}$ Hz.

    According to the remote observer, the astronaut on the Moon will therefore not only measure a higher bouncing frequency on the mirror (per point 2 above), but the incoming Photon also appears to contain a proportionally larger Photon impulse force.

    Based on these two accumulated effects (according two point 2 and 3), the apparent

repelling impulse force is expected to be a
quadratic *function of the* **'orbit diameter**
correction factor' *(that is: the length of the*
'remotely observed diameter', divided by the
length of the 'locally observed diameter').

4. (The repelling force) depends on the angle of
   impact of the Photon on the mirror:

   the larger this angle –relative to the direction in
   which the mirror is aimed- the smaller the
   resulting impulse force will be. This was referred
   to as the aforementioned 'impact angle effect'. In
   the extreme case the Photon comes in parallel to
   the mirror's surface, and the resulting impulse
   force is 0 (zero). The 'impact angle effect'
   dampens the quadratic effect of the
   aforementioned 'orbit diameter correction factor'.
   At relatively low orbit velocities –as in the current
   model- the effect may be considered marginal, but
   at the higher velocity cases (see the next chapter)
   it must be taken into account.

   The *apparent* (and *relative*) clock slowdown thus
   plays a key role in the model. Therefore it will
   now be quantified.

   Such quantification is possible because this
   slowdown is *entirely* based on two features:

   1. The velocity of Photons is finite, and equal
      from all perspectives, and,

2. The ratio between the locally and remotely observed object distances (LOD and ROD):

both distances thereby are based on the *equilibrium requirement* that is associated with a gravitational orbit. In this model the physical fact is that Moon and the Earth *do* orbit around their shared central point of mass. In any frame of reference…
To maintain this gravitational orbit, the requirement therefore is that within each frame of reference *the centripetal force and the gravity force must be in balance.* This requirement obviously will dictate a relationship between mass, the observed object distances (or: orbit diameters), the orbit velocities, and gravity.

Note that the above model illustrates that it does *not* need to be the shear fact that 'the Moon has a gravity field' that makes a local clock run at relatively slower pace. In Metric Physics the presence of a 'gravity field' is given as *the* cause for a –relative- clock slowdown near mass.

The above model of orbiting objects as well describes relativity of time measurements, but here through the already addressed properties and appearances that are part of Crenel Physics. This includes the equations that were verified in chapter 9. And this model will show –see the following- that it is the *combination* of the interlinked parameters that explains and quantifies the relative clock slowdown.

Note: in Crenel Physics –as was discussed earlier- some dynamics (or: movements) are required for creating 'some period of time' which is a requirement to allow detection of an object. Dynamics were a requirement for time measurements. The current model illustrates, that the relativity of these time measurements (in Metric Units) is also associated with dynamics (=velocities). This is not a coincidence. If an object can only be detected 'if something moves around' (this is the Crenel Physics equivalence of 'must exist for some period of time'), it is only logical that it can only be movement that impacts time measurements and may make these relativistic.

The objective thus is, to express the clock slowdown as a correction factor, to be applied to the remote clock (which serves as the base case) in order to find the local clock pace. Figure (15.5) describes the situation at hand.

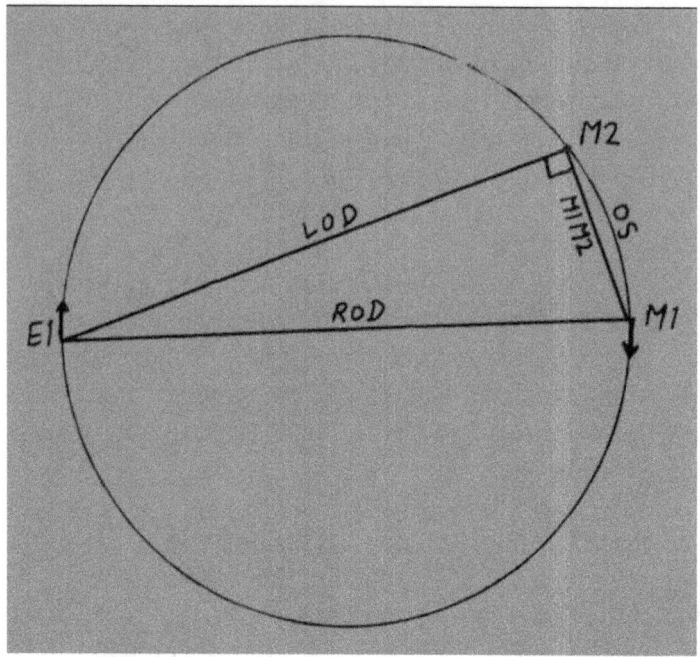

Figure 15.5: correction factor for observed object distances (or: orbit diameters).

According to Figure (15.5) the looked for correction factor is 'de facto' the ratio between the locally observed diameter LOD and the remotely observed diameter ROD. The looked for correction factor therefore is: LOD/ROD.

To quantify this correction factor, as the first step it is noted that point M2 may lie at a range of locations (pending the orbit dynamics), but in all cases it lies at some point *on* the shown orbit circle. At the same time, points E1 and M1 lie exactly opposite to each other on this circle. Therefore, irrespective to where point M2 may exactly be located, the angle E1-M2-M1 always is a

90^0 angle. The drawn triangle in figure (15.5) therefore is rectangular, and Pythagoras's theorem is applicable:

$$LOD^2 + (M1M2)^2 = ROD^2 \qquad (15.1)$$

From this equation the looked for correction factor LOD/ROD can be derived:

$$LOD^2/ROD^2 = 1 - (M1M2)^2/ROD^2$$

Or:

$$\frac{LOD}{ROD} = \sqrt{1 - \frac{(M1M2)^2}{ROD^2}} \qquad (15.2)$$

In equation (15.2) the term 'M1M2' represents the length of a straight line, per figure (15.5). The length of this line is associated with the length of the *actually* followed – and curved- orbit section which will be referred to as 'OS'.

The lengths of the orbit section OS and the orbit diameter ROD both are subject to the requirement for maintaining a stable gravitational orbit: the centripetal force F_{cp} and the force of gravity F_g must be equal to each other. The respective equations -quantifying these forces- were defined (and verified for Crenel Physics) in chapter 9.

Note that in the current model, the mass of the Earth and of the Moon were presumed equal. The required equations have been adjusted accordingly as follows:

$$F_g = \gamma.M^2/D^2 \qquad (15.3)$$

$$F_{cp} = M.\omega^2.r = 2.M.v^2/D \qquad (15.4)$$
(where v= ω.r and r=D/2)

Here, parameter 'M' represents the mass of the Earth and/or Moon, and 'D' the orbit diameter. From the requirement that $F_g=F_{cp}$ it can be derived from (15.3) and (15.4) that for a stable circular gravitational orbit:

$$v^2 = (\gamma.M^2/D^2) * (D/2M) = \gamma.M/2.D \text{ or:}$$

$$v= \sqrt{\frac{\gamma.M}{2.D}} \qquad (15.5)$$

This requirement for the orbit velocity 'v' can be used for calculating the length of 'M1M2' in equation (15.2) by considering the following:

1. According to the remote observer's clock, the required *timeframe* for sending a Photon from Earth to Moon is given by: ROD/c, where 'c' is the velocity of light, and ROD is the distance he sees between both objects.
2. According to the remote observer, the length of the orbit section 'OS' is given by the orbit velocity 'v' (per equation (15.5)), multiplied with the time that it requires a Photon to cross the orbit diameter (per point 1 above).

Thus: OS = 'v' x Photon crossing time.

This gives for the length of the curved orbit section 'OS':

238

$$OS = \sqrt{\frac{\gamma . M}{2. ROD}} \times \frac{ROD}{c}$$

$$= \sqrt{\frac{\gamma . M . ROD}{2 . c^2}} \qquad (15.6)$$

The length of the straight line M1M2 than is:

$$M1M2 = 2 . r . \sin\left(\frac{\theta}{2}\right)$$

$$= ROD \times \sin(\frac{\theta}{2}) \qquad (15.7)$$

In which 'θ' is the angle of orbit rotation relative to the centre of the orbit (and expressed in radials).

From the length of 'OS' and the diameter of the orbit, the angle 'θ' can be derived as follows:

$$OS = R \times \theta = \frac{ROD}{2} \times \theta$$

Or:

$$\theta = \frac{2 . OS}{ROD} \qquad (15.8)$$

The substitution of this result for θ into equation (15.7) gives:

$$M1M2 = ROD \times \sin(\frac{OS}{ROD}) \qquad (15.9)$$

When the result of Equation (15.6) is substituted in (15.9) the result is:

$$M1M2 = ROD \times \sin\left(\frac{\sqrt{\frac{\gamma.M.ROD}{2.c^2}}}{ROD}\right)$$

Or:

$$M1M2 = ROD \times \sin\sqrt{\frac{\gamma.M}{2.c^2.ROD}} \qquad (15.10)$$

This result for 'M1M2' can be substituted into equation (15.2):

$$\frac{LOD}{ROD} = \sqrt{1 - \frac{\left[ROD \times \sin\sqrt{\frac{\gamma.M}{2.c^2.ROD}}\right]^2}{ROD^2}} \qquad (15.11)$$

This can be rewritten as:

$$\frac{LOD}{ROD} = \sqrt{1 - \left[\sin\sqrt{\frac{\gamma.M}{2.c^2.ROD}}\right]^2} \qquad (15.12)$$

Or, because in goniometry a general equation based on Pythagoras exists:
$$sin^2(x) + cos^2(x) = 1$$
Which can be rewritten as:
$$\cos(x) = \sqrt{1 - sin^2(x)}$$

The above right term matches the right term in equation (15.2), which therefore can be written as follows:

$$\frac{LOD}{ROD} = \cos\sqrt{\frac{\gamma.M}{2.c^2.ROD}} \qquad (15.13)$$

Or, by using equation (5.5):

$$\frac{\mathbf{LOD}}{\mathbf{ROD}} = \cos\left(\frac{v}{c}\right) \qquad (15.14)$$

Equations (15.13) and (15.14) both give the clock slowdown factor on Earth or Moon, relative to the remote clock on the North Polestar. The equations are applicable to the current model where two equal objects orbit around a central point of mass.

The clock slowdown correction factor that is derived in Metric Physics through the theory of relativity can be found in literature (the theory of relativity):

$$\sqrt{1 - \frac{2.\gamma.M}{c^2.R}} \qquad (15.15)$$

Note, that this equation has some similarity with (15.12). Equation (15.15) –from Metric Physics– applies however to another scenario: it is the clock correction factor applicable to one single (non rotating) mass 'M', where variable 'R' represents the distance from this mass. As expected, at infinite distance there is no correction (or: the correction factor equals 1), while -as this distance to the mass decreases- the local clock will slow down relative to the remote clock. At a distance 'R' equal to $2.\gamma.M/c^2$, the local clock will have come to a full stop (relative to any remote clock). Because 'γ' and 'c' are physical constants, there is –for any value of 'M'- an associated value for distance 'R', at which the clock stops:

$$\frac{R}{M} = \frac{2 \cdot \gamma}{c^2}.$$

Equations (15.13) and (15.14) in turn give a clock slowdown applicable to the current model of gravitational orbits. The given correction factor is the clock slowdown factor *on* either one of the two orbiting objects, relative to a clock at a remote location. In the Earth/Moon model the local clock would come to a full stop when the orbit velocity equals $\pi/2$ times the speed of light 'c':

$$\cos(\frac{\pi}{2}) = 0.$$

This requirement obviously cannot be met, because it would require an orbit velocity *above* the speed of light.

Should an observer travel from the North Polestar to one of the orbiting objects, e.g. the Earth, his clock should – upon completion of the trip- be in pace with the Earth's clock. While approaching the Earth, the traveller would find it's clock *gradually* appear running at closer pace relative to the Earth's clock: one would not expect a sudden step change in clock pace between two adjacent points on the journey.

To explore this pace transition further, consider the following path: first travel from the North Polestar along a straight line to the centre of mass of the Earth/Moon system. From the followed reasoning it becomes clear, that none of the above calculations would thereby be impacted: the light coming from both the Moon and the Earth would still be equally retarded upon arrival at the observer's position. The initial 'Remotely observed orbit diameter' would still be seen, and remain unchanged. Therefore, the correction factor as found in Equation (15.13) (and (15.14)) is also valid at the central point of mass of the Earth/Moon system: the clock in this centre will run at the same pace as the clock on the North Polestar (or any other 'static' location). This seems in conflict with equation (15.5) from Metric Physics, according to which the clock at the central point of mass should run slower because one now is closer to the masses of both the Earth and the Moon. These masses are however orbiting, which can be seen as being in a continuous free fall (or acceleration) towards each other. This is the reason why equation (15.5) is not applicable here.

In the current model, the anticipated shift in clock pace therefore must entirely take place while travelling the second part of the journey: from the shared centre of mass (and rotation) towards one of the orbiting objects (Earth or Moon).

Out of the many possibilities, there are two scenarios for this second part of the trip that are of particular interest:

1. A 'radial' path from the centre of gravity towards the orbit.
   Here, the observer would travel along the straight line indicated in figure (15.1) as the 'Remotely observed diameter', and thereafter he would need to accelerate to orbit velocity, and,

2. A 'spiralling' path.
   Here, the remote observer would –upon departure from the central point of gravity- simultaneously

start orbiting this central point at an angle
velocity equal to the Earth's. Thus, his path
would orbit in a spiral towards the orbiting Earth.

Any other potential path scenario can be seen as a
combination of the above two scenarios.

The aforementioned 'radial' path initially does not bring
the observer onto the Earth yet: it brings the observer
only to a 'static' point on the Earth's orbit path. While
arrived here, he would –at regular intervals- see the Earth
(and Moon) passing by with their orbit velocity 'v'. One
question thereby is: to which orbit diameter should the
observer decide to go upon departure from the centre of
mass? There obviously are two possibilities: the
'Remotely observed diameter' (ROD), and the 'Locally
observed diameter' (LOD). Upon departure from the
central point of mass, the observer sees the orbit radius
corresponding with the ROD. Therefore, upon arrival at
the orbit path –based on the ROD- it will physically and
regularly be 'hit' by the orbiting Earth (and Moon) which
is at velocity 'v'. Thus there is no reason why the
assessment of the orbit diameter –being equal to the
ROD- should change 'en route' (assuming slow
travelling). The final step of this journey scenario now is,
to catch up with the orbiting object (Earth or Moon). To
achieve this, the observer has to accelerate its velocity
from 0 (zero) up to 'v' along the orbit, and thereafter
maintain a continuous lateral acceleration to stay in orbit.
These accelerations cost no effort (or energy) because the
observer is presumed weightless. As soon as the landing
on the orbiting object is achieved, parameter 'v' equals 0,
and according to equation (15.14) the relative clock

correction factor than becomes 1: the traveller's clock is now in exact pace with the Earth's clock.

And thus the observer now completed his journey towards the orbiting Earth, he is residing on it, and his clock will now be in pace with the Earth's (or Moon's) clock. At the same time, the observer will now find himself orbiting at the smaller LOD, rather than the ROD on which it resided prior to the acceleration. The observer now finally sees what the gunman has been seeing all the time.

According to equation (15.14) and given the model underneath, it ultimately was the *relative difference in velocities* 'v' (and no other variable) that caused the clocks to run at different pace. And it also was this difference in velocities that caused the orbiting radius to *appear* smaller.

One can only conclude that it physically is one and the *same* orbit diameter that was followed during this second part of the journey... but now –after the acceleration- this *same* diameter only *appears* shorter. There are *no* two separate diameters, there are *no* two orbits: there only is *one* orbit with *one* diameter. This diameter has however different *appearances, all depending on relative velocities*. The observer did not physically jump from one diameter (the ROD) into another (the LOD): it is only the *appearance* of this diameter that shifted from the perspective of the travelling observer.

There was no physical jump from one orbit into another.

The model illustrates the relationship between the change in clock pace, and the observed *compression* of distance measurements: while the velocity difference between observer and Earth was decreased from the initial 'v' m/s towards 0 m/s, the –now local observer- saw the apparent orbit diameter (expressed in meters) decrease from the initial ROD to the smaller LOD.

These are all considerations from a remote perspective: while travelling, an observer will never notice his own clock to slow down (or his meter getting shorter). It is always someone else's clock that appears to run slower, and someone else's meter to appear shrunken. The changes in appearances depend on the relative velocity - the parameter 'v'-, which in turn (in this model) was subject to the equilibrium associated with a gravitational orbit. Thereby, the direction of the velocity was not relevant: the effects are the same along the entire orbit path.

Had –instead of the 'radial' path- a 'spiralling' path been followed (the second scenario), the observer's velocity would have gradually increased as the radius of his orbit was increasing (per equation: $v=\omega.R$). The outcome upon landing on the object (Earth or Moon) would however have been the same after all.

At the bottom line it also is the appearance of the 'distance' yardstick (the meter) that depended on the chosen perspective.

The *number of wavelengths* of some Photon could have been used as an alternative to the 'meter', for expressing distances such as the various apparent diameters of the orbits ROD and LOD. Had distances been expressed in terms of a 'number of wavelengths' (of some arbitrarily selected Photon), this counted number (and/or fractions thereof) would have been equal to all, regardless any possible perspective. When expressed in number of wavelengths, there would have been *no apparent difference* between the 'ROD' and 'LOD'. Instead, the *frequency* of the particular Photon would have been judged differently, based on the differences in pace of the various clocks. And thereby the size of the wavelength (in meters) would be judged differently, because from any possible perspective the Photon travels at the same light speed.

In Crenel Physics, distance (and 'time') is expressed in Crenel, and thus there likewise is *no* difference in the numerical values for LOD and ROD. The aforementioned alternative for expressing distances in terms of 'number of Photon wavelengths' (rather than meters) therefore has the same impact, as describing it in the terms of Crenels (as per Crenel Physics). The Crenel Physics approach is therefore not as eccentric as it may have appeared at first sight.

In Crenel Physics, spatial sizes do not shrink, and clocks do not slow down. A Crenel is a Crenel, and a Package is a Package. The above model showed that the Package containment (of the crossing Photon) appears to increase, as the appearance of spatial 'sizes' (the orbit diameter and the crossing Photon's wavelength) is decreasing: Crenels are indeed reciprocal to Packages. Crenels are swapped with Packages. Thereby the product P.C (Packages x Crenels) contained within the model remains constant. This interpretation reflects the Crenel Physics version of the theorem of 'conservation of mass and/or energy':

Within a 'world', the product of contained P.C is constant:

P.C=constant

The difference with Metric Physics is the enhancement of the equation: an exchange between Crenels and Packages is possible. Furthermore, the definition of 'world' is not equal to the definition of an 'enclosed system' as used in Metric Physics: the concept 'enclosed' relates to spatial boundaries, and these have no meaning in Crenel Physics. Here, any object expands the appearance of its spatial sphere of influence (and e.g. its strength of gravity) at light speed, and 'containment' of this ever expanding appearance is not possible.

Because different *apparent* results for the orbit diameter were found when using metric units, each system de facto uses its own metric yardstick for measuring distances. Or more specific to the current model: de facto a 'meter' in the remote frame of reference is not equal to a 'meter' in the local frame of reference. From the remote perspective the 'LOD' appears shorter, and therefore the 'meter' in the local frame of reference appears shorter. Or: if the remote observer cuts of one meter of rope, and gives this rope to the gunman on Earth, this piece of rope will now appear shorter from the remote perspective. From the gunman's perspective however, he indeed received a meter of rope.

From the gunman's perspective, all information in previous figures is misleading. Here, the coordinates of all indicated points are based on the remote yardstick for distance measurements, while –according to the gunman- the local yardstick should have been used instead.

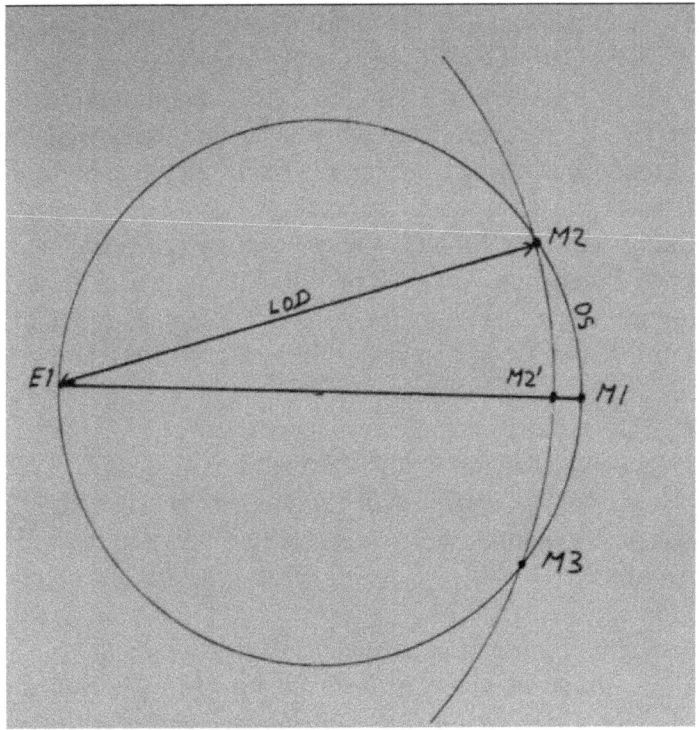

Figure 15.6: the Photon's perspective.

There also is the Photon's perspective on the model, see Figure (15.6).

Should the Photon -that was fired- have a rear-view mirror, in this mirror -and while underway- it would *not* notice the laser gun (or the Earth) move in its orbit. From its perspective, the Earth –from which it was fired- stays put at point E1. Through its front windshield however (see figure 15.6) it would reckon that the Moon is orbiting at double speed: when it was fired it saw the Moon at position M2, and upon arrival (about 1.3 seconds later) the Moon is in position M3. Both points M2 and M3 are at equal distance from E1. Thus, in these 1.3 seconds the Moon appears to have travelled an orbit section from M2 to M3 via point $M2^I$. The Photon thus 'possesses' its own private perspective and frame of reference, according to which the fixed reference point is the laser gun from which it was fired: point E1. According to the Photon's perspective, this reference point of origin has *not* been moving through space. And around this *fixed* and unique point of reference, this point of origin, the Moon appears orbiting at a *radius* equal to the 'LOD', or: at a diameter *double* the indicated 'LOD'. From the remote perspective –as per Figure (15.6)-, the associated *apparent* length of this orbit path does however appear *shorter* than 1500 meters because this orbit path via point $M2^I$ is less curved relative to the path via point M1.

> *Within the Photon's frame of reference, the apparent radius is the exact **double** of the radius that the gunman on the orbiting Earth sees (the locally observed diameter 'LOD').*
> *Simultaneously the Moon appears to be orbiting at double orbit velocity along this **double** radius.*

The Notohp.

In the above model a Photon was bouncing between two equal objects that are orbiting around their central point of gravity. The result was a *repelling* force between both objects. A likewise model can be applied to describe the functioning of an *attracting* force (and name it 'gravity force'). Albeit that instead of a repelling force being based on Photon impulse, an attracting force would need to be based on something else.

In Metric Physics, the -hypothetical- particle 'graviton' is introduced to describe gravity. A graviton is a hypothetical object, because it never has been found, nor isolated, nor identified.

There is a risk in introducing such a 'graviton' into Crenel Physics, because this may lead to ambiguity in terminology when properties are assigned. It is for this reason that -instead- a new Photon like type of elementary particle is introduced here:

> *Definition: the 'Notohp':*
> *A 'Notohp' has all properties of a Photon, except that its velocity is not +1, but its velocity is -1. Relative to a Photon, it 'moves' in the **opposite** direction of its velocity vector.*
>
> Note that the name 'Notohp' is the reverse spelling of the word 'Photon'. This helps memorising the name.

Moving in the opposite direction of the velocity vector cannot be envisioned within a spatial frame of reference. However, in Crenel Physics the process of 'movement' is nothing but a conversion between the 'time' appearance and the 'distance' appearance of the Crenel. Where a Photon –while moving- is converting the 'time' appearance of the Crenel into its 'distance' appearance in a one-on-one ratio, the Notohp is converting this same 'time' appearances into *negative* 'distance' appearances in a one-on-one ratio. Consequently, the impulse vector of a 'Notohp' points into the direction where it *came* from, whereas the impulse vector of a 'Photon' is pointing into the direction where it is going to. While Photons can be –partially- envisioned by the human brain as 'particles', such an envisioning of a Notohp movement is not possible. At first 'sight', Notohps should fall apart because their direction of movement is opposite to their impulse vector. The concept does not fit into the human way of *envisioning*. But nevertheless it can be *imagined*.

> Keep in mind that in Crenel Physics the human *spatial* envisioning –and also motion within such a spatial space- is nothing but an appearance. As colour seeing is nothing but an appearance… see chapter (13). A hypothetical *fourth* colour receptor in the human eye (e.g. for ultra violet, as some insects have) seems technically entirely possible, but it would require an extrapolated remodelling of the current colour seeing by humans. Such an extrapolated model now would require *four* spatial dimensions when laid out. Nevertheless, conceptually such an extrapolation with a fourth colour adapter is humanly conceivable albeit that the colour models become mathematical and abstract. These models can be *imagined*, but cannot be *envisioned* because humans do not see a fourth dimension in space.

> Likewise, the Notohp can be seen as an extrapolation (or perhaps even as a counterpart) of

the already introduced Photon, only conceivable through extrapolation of what *can* be envisioned.

Note that -by replacing Photons with Notohps in the above model- this model now describes the Crenel Physics mechanism for an *attract*-force, which will be named 'gravity force'.

In retrospect:

The Earth/Moon model used in this chapter was described in the terminology of the Metric Physics: here, spatial orbits can be graphically represented and envisioned, as in figure (15.1). Furthermore, time and distance were indeed –in line with Metric Physics- treated as two separate entities measured in meters and seconds respectively. The consequence was that there is no universally valid numerical result in expressing a distance –or time- anymore. The above model made that clear. In this model one has to deal with *two* (or more) different –apparent- orbit diameters for *one* physical orbit, and *two* (or more) –apparent- clock paces for *one* physical clock. The result of quantifying a distance (between two objects) or time (between two events) now depends on *where* one is (on the Moon or Earth, or the North Polestar, or elsewhere) and *how fast one is moving*.

These dependencies are reflected in the theory of relativity. A problem with the term 'theory of relativity' is that this is not a *theory*: it is fact.

In Crenel Physics the time/distance between Earth and Moon is measured in Crenels. And this number of

Crenels does *not* depend on where one is, or how one is moving. The model did also show a transfer from Crenels into Packages: the appearance of the longer ROD shrunk into the shorter appearance of the LOD, while simultaneously the Package appearance of the model components (the crossing Photon, the orbiting objects) increased:

> *The product of 'Packages' x 'Crenel' in the model thereby remained constant from any perspective (remote or local).*

> *This was the –broader- Crenel Physics interpretation of the principle of 'conservation of mass and/or energy'. In Crenel Physics the 'number of Packages' contained in a closed system is <u>not</u> constant: Packages can be exchanged with Crenels, pending the perspective of the observer. Furthermore, 'containment' is not possible because each object expands its spatial world of influence at light velocity. Gravity fields cannot be contained.*

> *The energy of the 'orbit crossing Photon' (expressed in Packages) appeared higher to the local observer, which's clock appeared running at lower pace (expressed in Crenel).*

The properties of a gravitational orbit (that is: its radius and the velocity within the orbit), and the perceptions of distances and times were analyzed. And because gravitational orbits *are* depending on mass and gravity, the above experiment illustrated how distances, time, masses and gravity indeed are related to each other. They therefore *cannot* be completely independent parameters.

The found relationship is fully embedded in Crenel Physics: here, time *is* distance (both measured in Crenel), and mass *is* set equal to 'Strength of Gravity' (both measured in Packages), and Crenels *are* reciprocal to Packages.

The question than is, whether something more can be learned from the experiment as illustrated in figure (15.1). This experiment described the perspective as one would see it from the North Polestar. What sneaked however into the experiment was the observation that from this position on the North Polestar, one saw Moon and Earth *rotate in an orbit* around their centre of gravity. If there would be just this single remote position on the North Polestar, and the Moon and the Earth, how could one possibly conclude that Moon and Earth are actually *orbiting* (in a perpendicular plane)? By human intuition, this orbiting would be envisioned relative to a larger and static universe around this system. But how could one detect this orbiting without having additional reference points within this larger encompassing universe? These additional reference points have not been mentioned in the above model.

It can be assumed in this experiment that these other reference points are there, but were *not* detected.

Without these other points of reference being detected, according to Metric Physics one could (and would) still have the possibility to conclude that there must be such a thing as *gravitational force*. Because Earth and Moon are not collapsing towards each other, there must be some counterforce that compensates this gravitational force. This counterforce was found in the centripetal force that is associated with a continuous tangential acceleration, or: with a rotational movement in an orbit.

Chapter 9 listed some formulas that are applicable here. The parameter 'ω' is used to quantify such a rotation. The question then is: *what* exactly is rotating? One could assume that the Earth and Moon system is rotating in their larger universe. But – in lack of reference points – one may as well argue that this larger universe is rotating around the Earth and Moon system.

In lack of additional detectable reference points, one could not possibly decide whether the Earth and Moon system is rotating in a stationary encompassing universe, or whether this encompassing universe is rotating around a stationary Earth and Moon system. Without additional points of reference, the concept 'stationary' has no anchor point. And the parameter ω has a floating – relative- value.

Regardless the choice of reference: in both options there is a process called *rotation*, which is –in Metric Physics-expressed in radials per unit of *time*. In Crenel Physics, rotation is expressed in radials per Crenel. Thereby, one full revolution corresponds to a rotation of $2.\pi$ radials. One full revolution thus corresponds with the full recuperation of spatial coordinates at the cost of some elapsing time.

In Metric Physics, rotation requires –at minimum- a 'plane' in which it can take place: that is a 2-dimensional spatial space.

In Crenel Physics however, rotation may also take place in a 1-dimensional space. As was described earlier (see chapter 6, table 6.1), from a 2-dimensional perspective such a 1-dimensional spatial space looks like a series of *concentric circles*. And rotation can take place along such a circle.

From a 3-dimensional perspective, such a 1-dimensional space looks like a series of *concentric spheres*. And rotation can take place along such a sphere. And so on. Obviously, the rotation cannot be *detected* within the 1-dimensional space itself, but it can be detected from the 2-dimensional (and 3-dimensional, 4-dimensional, etc.) perspective.

This leads to new/enhanced definitions in Crenel Physics:

> *Movement:* the conversion of the time appearance of the Crenel into one or more spatial appearances of the Crenel (or vice versa).

> *Rotation:* the conversion of the time appearance of the Crenel, *perpendicular* to a spatial appearance of the Crenel. To detect rotation, one therefore needs at least one extra spatial dimension of the Crenel.

Any path can be described as a combination of the aforementioned *movement* and *rotation*. Both *movement* as well as *rotation* are relative to some point of reference.

An experiment was discussed to detect *rotation*:

> If *two* Photons are fired –rapidly after each other-
> into a direction where a target is *seen*, and if
> thereafter *both* Photons hit the target, there is *no*
> rotation. If one or both Photons miss the target,
> the conclusion is that the target has a tangential
> velocity relative to the gunman.

16. Orbits at high speed.

This chapter explores some more details of higher orbit velocities. Therefore, the earlier introduced 'Earth/Moon model' will be re-used, with the same simplifying assumptions:

1. the Earth and the Moon are of equal mass, and
2. very small in spatial size,
3. The location of the –remote- observer (on the 'North Polestar') is perpendicular to the plane of rotation, and straight above the shared centre of mass.
4. There still is a 'gunman' on Earth, an 'Astronaut' on the Moon, and a 'remote observer' on the North Polestar.

As seen from the North Polestar, Earth and Moon appeared orbiting at a radius indicated as the 'Remotely observed diameter' ('ROD'). From the local Earth or Moon perspective the orbits were found to have a shorter diameter referred to as 'Locally observed Diameter ('LOD').

The difference between ROD and LOD was caused by the usage of Metric units that are *relative* units of measurement. Had the orbit distance been expressed in e.g. a number of wavelengths from some arbitrary Photon, or in Crenels, that number would have been equal from both the remote as well as the local perspective.

Figure 16.1 redraws the earlier used model from the perspective of the remote observer, thereby showing the impact of potentially higher orbit velocities.

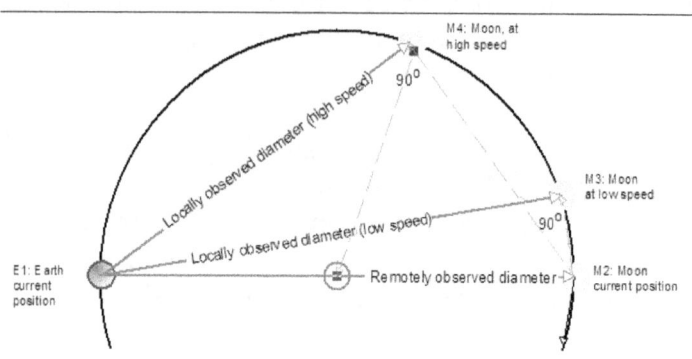

Figure 16.1: the impact of orbit speed on the observed orbit diameter.

When the Moon travels very slow relative to the velocity of light (which it indeed does...), according to the remote observer, the gunman on Earth could see the Moon e.g. at position M3. That is a position that the Moon had some time ago. In the figure this point M3 is drawn out of scale for graphical reasons: the distance between M2 and M3 would be approximately 750 meters, while the distance between E1 and M2 would be approximately 400,000,000 meters.

As the figure also illustrates, the distance between E1 and M2 (the ROD, or: 'Remotely observed diameter') is indeed longer than the distance between E1 and M3 (the LOD, or: 'Locally observed diameter').

There is a second aspect to this: the gunman will always see the Moon at the exact opposite point in his own orbit. That is a position straight behind the shared centre of mass that lies in between them. Therefore, figure (16.1) is somewhat misleading: it suggests that the gunman sees the Moon in the *direction* of point M3. In reality however, the gunman sees the Moon in the direction which is exactly opposite of its own position in his orbit, straight behind their shared centre of mass that supposedly is not moving. This shared centre of mass has fixed coordinates, both in the remote frame of reference as well as in the local frame of reference.

Figure (16.1) also illustrates that if *higher* orbit velocities are assumed, the difference between where the Moon *is* (from a remote perspective) and at what distance it actually will *appear* to an observer on Earth, will increase. Point M4 illustrates this.

According to equation (15.5) from the previous chapter…

$$v = \sqrt{\frac{\gamma.M}{2.D}} \qquad (16.1)$$

…these higher orbit velocities can –amongst others- be expected at shorter orbit diameters. The here shown equation comes forth from the requirement of equilibrium between repelling centripetal force and attracting gravitational force.

Within this context two model properties were addressed, relevant to Figure (16.1):

Property 1:
According to the remote observer, the
'retardation time' that an Earthly observer (at
position E1) has to deal with, is equal to the time
it takes for a Photon to travel from the Moon
(position M2)towards the Earth (position E1).

The reasoning behind this is that the duration of the 'retardation time' is equal to what a remote timekeeper - residing on the North Polestar- will reckon it to be: he indeed sees the orbit as indicated in Figure (16.1) along the 'Remotely observed diameter' ('ROD'). Albeit that he sees this with some –fixed- time delay. And based on this ROD he will calculate a 'retardation time' that would be applicable to an Earthly (or: local) observer. This calculated 'retardation time' would be measured on his (remote) clock.

Note that the gunman on Earth and the astronaut on the Moon will indeed see the shorter orbit diameter LOD. But at the same time their clocks are running proportionally slower relative to the remote clock.

The second model property is based on the physical fact that light can only originate from *actual* objects. Physical light cannot be emitted from a previous position of the Moon (although this so appears to the Earthly observer).

Property 2:
An Earthly observer will see a retarded Moon
position that <u>must</u> lie on the orbit path of the
Moon. The remote observer will find this position
by backtracking along the orbit path as seen.

As was discussed in the previous chapter, the question here is: what is the 'actual' orbit path that the Moon is following? It was found that – when distances are expressed in meters - there is no universally shared size of the diameter of the orbit... so what then is meant with 'actual'? Here, whatever a remote observer sees is taken –by choice- as the reference for 'actual'. This is consistent with the approach taken in the previous chapter.

To the remote observer on the North Polestar, both Earth and Moon do appear to follow the orbit as drawn in Figure (16.1). Given the –constant- retardation time associated with the remoteness of his position, and given the orbit speed of the Moon, he can now backtrack from the –reckoned- current Moon position in a reversed orbit direction along this past orbit path, to end up in e.g. point M3, or M4. Based on the above two model properties the remote observer can thus reason at what exact distance the Earth based gunman will see the Moon (and the Moon based Astronaut will see the Earth).

Figure (16.1) illustrates that the higher the velocity in orbit (parameter 'v'), the closer the apparent Moon position (as seen from Earth) will approach the actual Earth position.

The actual orbit velocity in a stable orbit is based on the force of gravity being equal to the centripetal force, as indicated by equation (15.5). Despite this rule for equilibrium, there is also a limit to keep in mind: the Moon/Earth cannot travel faster than light. This is not very relevant for objects like the Moon and the Earth, as both have velocities that are only fractions of the velocity of light.

To explore this limit, a likewise thought experiment can be envisioned using Photons as orbiting objects. Photons by definition *do* have the velocity of light. Photons also are containers of Packages, and therefore have a 'mass' appearance in Metric Physics. They therefore will interact through their mutual 'Strengths of Gravity' and would fit into the model as the Earth and Moon do.

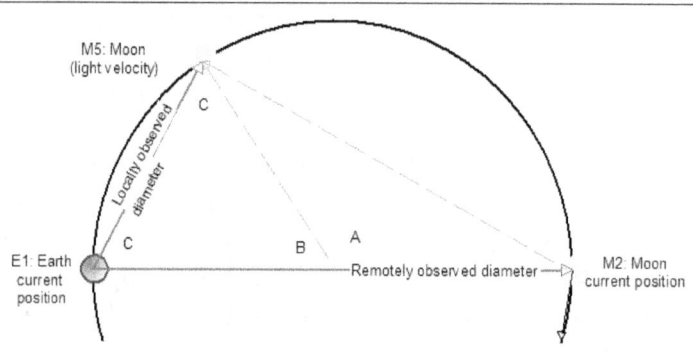

Figure 16.2: orbiting at light velocity.

Figure (16.2) illustrates the case: it thereby shows the maximum possible difference between a remotely observed diameter ROD of an orbit, and its -smaller- locally observed diameter LOD. Consider for this case that the words 'Earth' and 'Moon' are just names for two separate Photons, rather than the names of two planetary objects.

The length of the 'backtrack' path (the distance between point M2 and point M5) along the orbit path would now be exactly equal to the diameter of the orbit. This is based on the aforementioned 'Property 1' and 'Property 2'.

Because the length of the entire orbit circle is equal to $\pi.D$, thus representing a full revolution (which is an angle turn of $2.\pi$ radials), the angle marked 'A' in Figure (16.2) equals $(D/\pi.D) \times 2.\pi = 2$ radials. Angle 'B' then equals $(\pi - 2)$ radials.

The length of the 'locally observed diameter' then equals (based on goniometry):

$$LOD = ROD \times \sin\left(\frac{\pi-2}{2}\right) \qquad (16.2)$$

The numerical value of this sinus is approximately 0.5403, and therefore the locally observed radius LOD can never be less than this 0.5403 times the remotely observed radius ROD.

This result is consistent with the earlier derived equation (15.14) from the previous chapter…

$$\frac{LOD}{ROD} = \cos\left(\frac{v}{c}\right) \qquad (16.3)$$

… if in this equation (16.3) for velocity 'v' the – maximum- light velocity 'c' is substituted: the value for the LOD/ROD ratio then is $\cos(c/c) = \cos(1)$.

According to goniometric rules, it is indeed so that:

$$\cos(1) = \sin\left(\frac{\pi-2}{2}\right) \approx 0.5403 \qquad (16.4)$$

Therefore, the earlier found equation (15.14) confirms the result of the model at light spced, as shown in figure (16.2).

There is however an important difference in the whereabouts of both results:

- To arrive at the result of the previous chapter, the properties of a circular gravitational orbit were used to derive the equation: the gravitational force F_g balances the centripetal force F_{cp}. This requirement is expressed by equation (16.1)… which was copied from equation (15.5).
- In the here followed approach where the orbiting objects are presumed Photons, the actual size of the orbit diameter is *irrelevant* to the result of the found ratio between LOD and ROD.

Also, the 'light speed' variant of the model contains an interesting relationship with the fundamental units of measurement from Crenel Physics. It can be found by following the approach of the previous chapter (involving gravitational properties), and thereby substituting light velocity 'c' for the orbit velocity 'v'. This would work as follows:

Take the earlier used equation (7.1) from Metric Physics:

$$E = m.c^2 = h.\upsilon \qquad (16.5)$$

From this, it can be derived that…

$$m = \frac{h.\upsilon}{c^2} \qquad (16.6)$$

…in which the frequency 'υ' can be written as v/2.π.r. Here parameter 'v' is the velocity of the object. This substitution in (16.6) gives:

$$m = \frac{h}{c^2} \times \frac{v}{2.\pi.r} = \frac{h.v}{\pi.ROD.c^2} \qquad (16.7)$$

With an orbit velocity equal to 'c' this gives for a Photon's mass:

$$m_{(Photon)} = \frac{h}{\pi.ROD.c} \quad (\text{in kg}) \qquad (16.8)$$

Substituting this result in equation (16.1) whereby velocity 'v' equals 'c' gives:

$$v = c = \sqrt{\frac{\gamma.M}{2.D}} = \sqrt{\frac{\gamma.h}{2.\pi.ROD.D.c^2}} \qquad (16.9)$$

or (because 'D' and 'ROD' both represent the same variable:

$$c = \sqrt{\frac{\gamma.h}{2.\pi.ROD^2.c^2}} \qquad (16.10)$$

In equation (16.10) the only unknown parameter is the 'ROD', which can now be extracted:

$$ROD_{(2\ Photons)} = \sqrt{\frac{1}{2\pi}} \times \sqrt{\frac{\gamma.h}{c^3}} \ (\text{in meters}) \qquad (16.11)$$

The factor $\sqrt{\dfrac{\gamma.h}{c^3}}$ in equation (16.11) corresponds to the length of 1 Crenel, as derived in chapter 8 (see equation (8.9)):

$$\text{ROD}_{(2\,\text{Photons})} = \sqrt{\frac{1}{2\pi}} \qquad \text{(in Crenels)} \qquad (16.12)$$

This corresponds with approximately 1.6×10^{-35} meters. This is the remarkable relationship that was looked for: in the units of Crenel Physics per equation (16.12) the found orbit diameter is entirely based on a *mathematical* constant 'π', and on nothing else. The fundamental *physical* constants do not play a role here… This is not so when Metric units of measurement are used, as per equation (16.11).

It should thereby be kept in mind that the current model is based on *two* equal Photons, sharing one single orbit (the 'Earth/Moon' model) that takes place in a 2-dimensional plane.

According to the Crenel Physics units of measurement, the Package containment of each Photon is the same as the frequency at which the orbits are completed. The length of the orbit equals $\pi.\text{ROD} = \sqrt{\dfrac{\pi}{2}}$, while the orbit velocity equals 1. Thus, in the 2 Photon case at hand, the Package containment of each of the two Photons equals $\sqrt{\dfrac{2}{\pi}}$ Packages.

This is quite heavy relative to the weight of atomic particles: one Package corresponds to 5.46×10^{-8} kg or 5.46×10^{-2} mg (see chapter 8, equation 8.14), or 7.4×10^{42} Hz. Atomic nuclei are in the range of 10^{20} Hz...

One question therefore is, whether the found orbit radius would be the *only* radius at which the envisioned Photon pair could orbit in a stable 'gravitational' manner.

Furthermore, one may as well have assumed a model where Photons bounce back and forth along a 1-dimensional string of line. At first sight the thus resulting movement could potentially be envisioned as harmonic swing of a pendulum with accelerating and decelerating object velocities. In the case of a Photon that viewpoint would however not work, because there *is* no acceleration or deceleration possible when it comes to Photons: in all cases and under any circumstances Photons are presumed to maintain a constant velocity... the velocity of Photons. The full impact of such a 1-dimensional re-modelling will need to be investigated later.

17. Orbit energy.

Consider a Photon passing by. By definition (and trivially) it has the velocity of a Photon, regardless the frame of reference from which the observation takes place. In Crenel Physics, the concept of *velocity* is defined as: the conversion rate of the 'time' appearance of the Crenel into the 'distance' appearance of the Crenel. Velocity thus is a dimensionless number. And within any frame of reference the Photon's velocity is normalized to a numerical value of 1. Thus, photons by definition execute the conversion at the -maximum possible- rate of 1: one Crenel of 'time' converts to one Crenel of 'distance'.

The Photon velocity is a *natural* reference for a system of units of measurement. Such a system is needed to describe a process. Giving this photon velocity the numerical value of 1 –regardless the frame of reference- is by choice: it makes conversion between the 'time' and the 'distance' appearance easy.

The three key reasons why the Photon velocity can indeed be used as natural reference are:

1. *Photons are separate entities.*
   Or: one can monitor individual Photons.
2. *Photons do not overtake each other.*
   Or: they all travel at equal speed through any system.
   The Package containment per Photon is irrelevant to this velocity.

3. *Photons positively have a velocity.*
   Or: without velocity, Photons would seize to be.
   If one detects a Photon, it *has* a velocity.
   In metric Physics this fact converts to the
   formulation: the 'rest mass' of a Photon is zero.

The earlier figure (14.1) shows that from a spatial
viewpoint Photons are continuously residing on their own
'birth line'. They continuously reside 'at the edge' of
their own world of influence. One therefore cannot *sense*
a Photon coming, until it physically reaches the sensor.
Prior to that, one can only anticipate –based on the
Photons known properties- that a Photon exists, that it is
coming, and when it will arrive. For example: when on
the Moon a light is switched on, it can be anticipated that
thereby Photons are produced, and that about 1.3 seconds
later these will indeed reach the eyes on Earth. There is
however no *sensor* that could inform the eyes in advance
that a Photon impact is about to happen.

The Cylinder Model.

In chapter 7, a Photon was thought to be trapped in an
imaginary cylindrical container with a mirroring surface
at the inside. Within this container the Photon was
bouncing back and forth at Photon speed, while the
container itself could be held at rest in some spatial frame
of reference. This model matched the actual behaviour of
the appearance 'mass'. The shorter the cylindrical
container, the higher the internal bouncing frequency of
the Photon, and thereby the associated Package content
(per Planck's equation). Note, that thereby an object
containing 1 Package has been defined as an object with
a 'frequency' of 1 cycle per Crenel. Thus, an object with

a containment of 1 Package would be the equivalent of one Photon bouncing in a cylindrical container with the length of half of one Crenel. In such a container, after 1 Crenel of 'time' a full back and forth cycle is completed. This model expressed a reciprocal relationship between contained Packages, and Crenels representing the size of the imaginary container. This 'cylindrical container model' therefore was consistent with earlier findings.

The 'Cylinder Model' connects the immaterial character that Photons have (Photons themselves cannot be grabbed and held still) with the material world of matter (that can be grabbed while at rest).

It represents Planck's equation: $E=h.\upsilon$

The Circular Orbit Model.

The question is where such an imaginary container could possibly come from. This is where the earlier discussed Moon/Earth orbiting model (of the previous chapter) falls into place. That is: the high speed variant thereof, in which the orbiting objects are replaced by Photons. It is then 'gravity' that shapes the orbit, and thereby a spatial containment within some frame of reference. Here, 'gravity' is the one and only reason for containment (and thereby tangible mass) to exist. Albeit that in this model the Photons were contained in a 2-dimensional orbit rather than a 1-dimensional cylinder.

The earlier conclusion that 'Strength of Gravity' must be the equivalent of 'Packages' (and thereby concrete objects) fits. Furthermore, it now is logical to postulate that *if* this 'Strength of Gravity' is *the* mechanism to contain Photons (in an orbit), that it is this same mechanism that can be expected to exist between *pairs* of these containers.

> *The gravitational constant (in the Metrical Physics sense) thus is not only causing interaction between pairs of masses: it is the reason why these masses exist in the first place.*
>
> *In Crenel Physics terminology the 'Strength of Gravity' is just one appearance of Packages. 'Strength of Gravity' comes together with Packages. As 'mass' and 'energy' do (and perhaps other appearances such as 'bits of information', or' degrees of freedom'). This is an unambiguous way of anchoring the relationships.*
>
> *We **are** gravity. Our tangible world **is** built of gravity. Gravity is the 'any-thing' as shown at the basis of the earlier Figure 12.1 (representing 'a basic model').*
>
> *A better wording for 'gravity' would perhaps be: 'the possibility to interact'. Without this sheer possibility, tangible material objects themselves cannot exist. From this viewpoint, stating that 'gravity' is just a force between masses would be an understatement. Indeed this force exists and can be measured. But it is the hart of the matter that reveals this force. And this matter **itself** is gravity.*

One complexity of the orbit model as shown earlier in figure (16.2) is, that it requires two Photons to join together to keep each other contained in a spatial orbit (or: in a 2-dimensional confined space). Another is that at equilibrium this would result in quite a heavy piece of mass (in the order of magnitude of one Package).

This is an incentive to search for a simpler containment model involving just *one* single Photon. For this model one has to go back to the normalised Photon properties. Photons indeed reside at the edge of their own 'birth line' per figure (14.1). If a Photon is thought to be produced at some unique spatial point of origin (e.g. a laser gun), it sees this point of origin in its rear-view mirror as a spatially fixed point (see also earlier figure 15.6). From a spatial viewpoint this non-moving point can serve as reference point, but this birthplace also has a broader meaning:

> If the eye of an observer is hit by a Photon, at that moment the observer will in effect 'see' the birth of this Photon *at its place of origin*, and *not* at the place where the Photon actually *is* (that would be: in the eye of the observer).

> This –in part- explains why humans actually and always see light emitting objects at some distance. If humans would see Photons where they actually are, that is *in* the eye, it would not be possible to compose the directional information that is so important for the human spatial view of the world.

Note that the same finding with regard to the place of origin applies to *gravity* forces: these forces also appear to come from the point where they are generated and *not* from the point of impact at some gravity sensor. In this respect the propagation of a gravity force through spatial space behaves equal to the propagation of Photons. The earlier introduced 'Notohp' as a carrier of interaction between pairs of objects fits into this picture: a Notohp only differs from a Photon in that it has a negative velocity vector (or: it converts 'distance' into 'time', rather than 'time' into 'distance' as a Photon does).

Also note that this 'birthplace observation' differs from the Newtonian impact of forces that are engaged on an object: for Newtonian forces the origin is totally irrelevant. What matters here is the strength of the force, the direction of the force, and the point of engagement *at* the object. The location of the source of the Newtonian force plays no role in how it is impacting.

The Photon –while on the move- is separating itself at light speed from its point of origin, and therefore it will not 'feel' the presence of its birth place. Signals from there cannot catch up. However, as soon as the Photon hits a target like the human eye, this target will 'see' the birth of the Photon at its place of origin. And at that same moment –at the target- the Photon seizes to be what it was before. Through the interaction with the target, this target may e.g. completely absorb the Photon, thereby gaining the Packages that were contained by it.

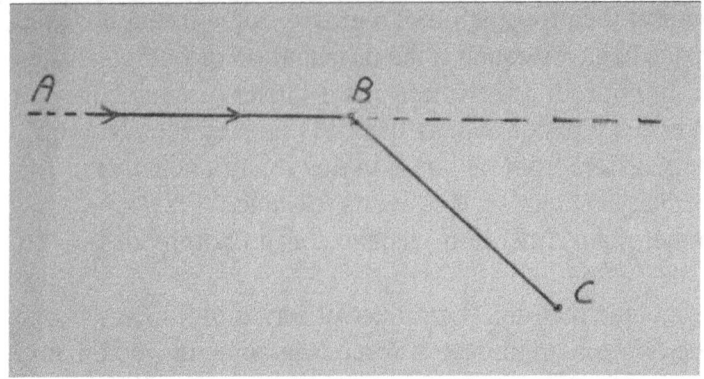
Figure 17.1: single Photon with bended path.

This viewpoint also has impact if –from a 2-dimensional perspective- the path of a Photon would be bent at some point, e.g. through interaction with some other object.

Figure 17.1 represents such a bended path. Here, a Photon originated from an unspecified remote point of birth 'A'. At point 'B' this path is –for some unspecified reason- bended towards point 'C'. The propagating Photon -after this path bending- now seems to originate from this bending point 'B', which apparently is a new birthplace.

This indicates that the original Photon originating from point 'A' seized to be, and that instead a new 'child' Photon is born at 'B'. It will be named 'child' here, because it originates from the initial Photon that served as 'parent'.

As said: the justification for this viewpoint of 'death' of the original Photon, and 'birth' of a new Photon is, that a sensor like the human eye –when hit by the child Photon-would indeed judge this post collision Photon to originate from point 'B', and not from any point prior to that.

And again, this new Photon itself will not 'feel' its unique birthplace 'B', because it is distancing itself at light speed from it.

> Note: this viewpoint of 'death' and new 'birth' is unlike the collision of a tangible object like a billiard ball: after the collision it is still presumed to be the same ball.

In this model, free travelling Photons are extremely glimpse objects relative to material –tangible- objects. As individual entities, free Photons *can* only travel straight. In fact, in Crenel Physics the path of a free travelling Photon has been used as the *definition* of 'straight'. Therefore the envisioning of the physics behind a course change is:

> A course change makes the original 'parent Photon' disappear while an entirely new 'child Photon' is born at its own unique birthplace. Both parent and child follow 'straight paths' during their lifetimes.

Figure (17.2) shows a *second* path bending at some point 'C', towards a point 'D'. Again the same reasoning can be followed. The 'child' that was born at point 'B' now seizes to be at point 'C', where a 'grandchild' is born that now moves into the direction of point 'D'.

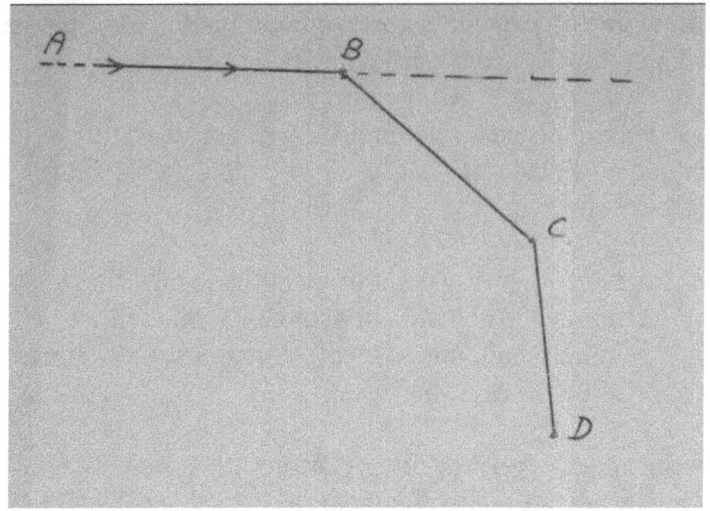

Figure 17.2: a second path bending at point 'C'.

Thus, in the model one now has some 'parent' that was born at some unknown remote point A, its 'child' being born at point B, and a 'grandchild' being born at point C. And none of these three Photons can interact with its own private birthplace, because each is distancing itself at light speed from it.

However, as the figure shows, the 'grandchild' is not distancing itself from point B at light speed… At the moment it is born at point C, it would –due to its course towards D- distance itself at a slightly *lower* speed from point B. Therefore, the 'grandchild' -immediately upon its own birth- will 'witness' the birth of its own parent at point B.

The impact of this is illustrated in figure (17.3).

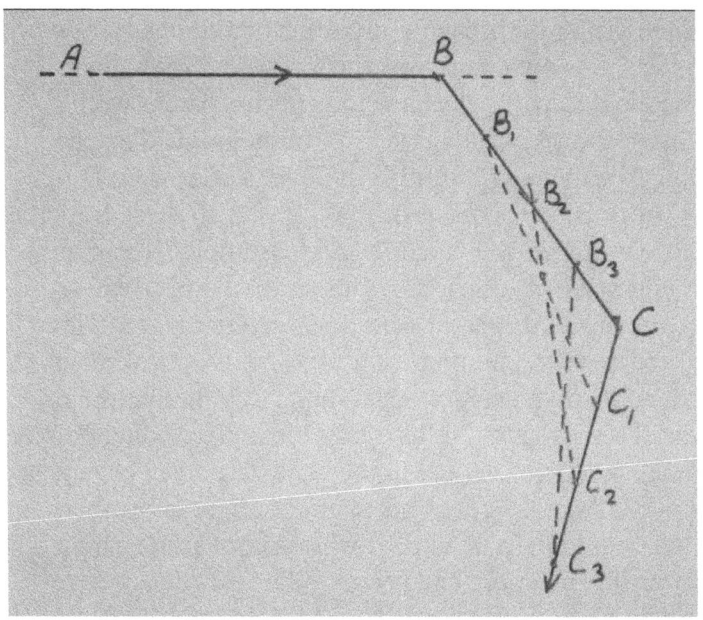

Figure 17.3: the impact of a second course change at 'C'.

When the grandchild is born at point C and commences its path in a new direction, it will simultaneously see the birth of its own parent at point B. When –for now- assuming no interaction, the dynamics look as follows: while the grandchild commences travelling from C to point C_1, it will 'see' its parent simultaneously travel from B to B_1. The path length from C to C_1 is equal to the path length from B to B_1 because both travel at equal (Photon) speed. Likewise, point C_2 corresponds to B_2, and C_3 to B_3. Until the grandchild would see its parent arrive at point C where it seizes to be…

During this initial episode of its own life (and still assuming no interaction with its parent) the grandchild 'sees' the entire life path –from birth until death- of its parent in its own rear-view mirror. And during this episode the spatial distance towards his parent would remain approximately constant. Initially this distance would be the length from B to C, thereafter from B_1 to C_1, from B_2 to C_2, from B_3 to C_3… and all these lines are –approximately- of equal length (assuming a relatively small course change at C). Therefore, the net relative movement of its parent would be seen in the rear-view mirror as a section of an –almost- *circular* movement around its own position: the grandchild Photon initially would see its parent in his rear-view mirror being born at some angle that is equal to the angle of course change at point C. And as he sees his parent's life pass by, this angle would drop to zero, at which moment its parent is not visible anymore and seizes to be.

This scenario becomes meaningful if one considers the following:

1. Because the parent Photon is indeed 'seen', there *will* be –other than assumed above- interaction between the Photon before a course change (the 'parent') and the Photon after a course change (the 'child'),

2. Any sizable course change can be seen as an integration of an infinite number of infinitely small course changes.

Both considerations combined will cause that, once a 'free' Photon is deviating from its originally straight path, the above scenario starts enrolling. From this moment onwards the Photon will pedigree from generation into generation in infinitely rapid successions, where each descendant *can* feel an interaction with its immediate parent.

This scenario raises a 'chicken and egg' question: any directional course change of any object can be associated with –at least part of- an orbit. And per Planck's theorem this –in turn- is associated with Package containment. And thereby there would be 'Strength of gravity'. Between pairs of objects this would then in turn cause exclusive interactions between pairs of objects, here parent/child interaction. The 'chicken and egg' point is that one needs a course change for such interaction to take place, and one needs interaction for the course change to be maintained.

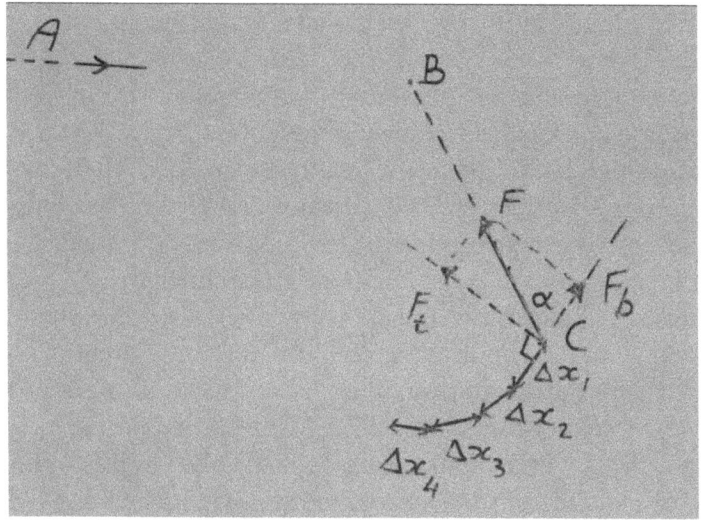

Figure 17.4: Remote perspective: an infinite number of infinitely small course changes.

Figure (17.4) shows what these parent/child successions look like when seen *from a remote perspective*. Assume that -for some unknown reason- at point C a course change of angle 'α' takes place. Thus a 'child' Photon is born at C. Immediately -at that same moment- there will be Package containment and the child will experience a gravitational force 'F' in the direction of point B, as indicated in the figure. More specific: a force pointed into the opposite direction of the course prior to the change. This force 'F' is composed of two components: a tangential component 'F_t' and a backward component 'F_b'. The force component 'F_t' is equal to $F.\sin(\alpha)$ and the component 'F_b' is equal to $F.\cos(\alpha)$.

The tangential component would cause the newborn Photon to change course, and thus result in the scenario of continued infinitely quick generation changes.

To explore what happens after the initial course change, assume that each subsequent generation thereby covers a very small path distance of length Δx Crenels, where Δx approaches value zero. This means, that the parent of any successive generation is also initially 'seen' at a distance Δx, and at some very small angle $\Delta \alpha$. While any successive generation covers its own distance Δx (which is equal to its own lifetime Δx), it will see its parent in the rear view mirror at an almost continuous distance Δx while the initial small angle $\Delta \alpha$ gradually reduces to value 0, at which point the parent's visibility seizes to be. Note that between the start of the angle reduction and the completion thereof (or: while underway) the seen distance between parent and child will actually be –only very slightly- less than Δx. The angle $\Delta \alpha$ becoming 0 marks the moment where the currently chosen generation itself also seizes to be, due to the start of a next infinitesimal course change. And the whole cycle starts over again with the next generation.

In this scenario it is now relevant to analyze the dynamics of the interaction between parent and child during their short lifetimes.

The first conceptual question thereby is: from what frame of reference is this analysis to be done? Figure 17.4 shows some –randomly chosen- *remote* perspective. Another and alternative frame of reference would be to position oneself locally *on* one of the Photons. Such a frame of reference would however be in conflict with the principle that Photons have the speed of Photons, regardless the chosen frame of reference. Should one position oneself *on* one Photon, one would violate that principle. It inherently will not be possible to take such a position. Therefore it is postulated here that remoteness is the only option. When it comes to Photons, only a remote frame of reference can be relevant to the outside world. And within such a remote frame, any Photon –by definition- travels at Photon speed.

From this remote perspective –as discussed- the child at its own birth will 'see' the birth of its parent. When spatially exploring its encompassing world, it is not only looking into spatial distances, but it is thereby also looking back in time. As discussed in chapter 14, key for visibility of an object is that the distance at which it resides (expressed in Crenels) is less than (or equal to) the age of this object (in Crenels). The distance at which the parent is born does exactly coincide with 'how far back in time' this happened. Or: the child is born on the birth line of the parent's world, and thereafter it will remain within this world because the parent's world is expanding slightly faster than the child Photon is effectively running away from it.

The next question is, whether -in turn- the parent could 'see' its child during its own lifetime. The answer is negative. To explain this, one has to view the coordinates of the parent's birthplace. The child is born at some distance Δx from these coordinates, and therefore it will take a timeframe of length Δx before this birth could possibly and initially be noticed by the parent. After this exact timeframe, the parent however seizes to be.

Because the parent does never see its child, it will experience no interaction whatsoever with it. Based on the laws of conservation of 'Package x Crenels' content (or in Metric Physics: based on conservation of mass/energy) it can then be concluded that:

> *Whatever* the child Photon may experience from the interaction with its parent, *it must have a net value of 0.*

To analyze this impact on the child, the two different impacts of the two force components 'F_b' and 'F_t' as shown in Figure (17.4) can be reviewed separately:

'F_b':

Force component 'F_b' would –on material objects and based on Newtonian rules- result in reducing the forward velocity. Photons however by definition do not obey these rules and cannot be slowed down.

There is nevertheless an amount of energy that *will* be transferred by this force component. This amount of energy is calculated as the strength of 'F_b', multiplied with the length of the path along which it is applied ($=\Delta x$).
Based on the same aforementioned energy conservation laws, the associated impact must result in an energy *reduction* of the Package containment within the Photon itself. This Package reduction is equal to: $F_b \times \Delta x$ or:

$$\int_{\alpha=0}^{\alpha=\alpha} \cos(\alpha) \times \Delta x \times F \qquad (17.1)$$

This can be rewritten as:

$$\sin(\alpha) \times \Delta x \times F \qquad (17.2)$$

In this equation parameter 'F' represents the gravitational force between the parent Photon and the child Photon.

Because force component $\mathbf{F_t} = F.\sin(\alpha)$, the reduction of package containment due to the force component $\mathbf{F_g}$ of the child Photon can be rewritten as:

$$\Delta x \times F_t \qquad (17.3)$$

This is the amount of Package *reduction* per small orbit section of length Δx that is caused by force component $\mathbf{F_b}$.

'Ft':

As indicated, based on conservation of energy the Package *reduction* expressed by equation (17.3) must match the Package *gain* that is associated with force component F_t . At the current micro scale of just one generation, this force component results in a tangential acceleration. However, because this tangential acceleration is inherited from generation to generation, it ultimately leads to the appearance of a circular orbit. In fact it is not one Photon that is orbiting, but a sequence of successive Photon generations.

This orbit brings a new parameter into the model, namely the apparent orbit radius 'R'. It is not a 'real' orbit radius because nothing is really orbiting: instead a whole series of successive Photon generations together shape a circular path.

The parameter Δx in equation (17.3) can than be rewritten as $R.d\alpha$, where $d\alpha$ represents the size of the infinitesimal small course change of the child Photon relative to its parent:

$$\Delta x \; = \; R.\, d\alpha \qquad\qquad (17.4)$$

Furthermore, the 'appearance' of the Package called 'mass' (symbol 'm') can now be introduced and quantified. It appears as a parameter into the following equation that is –generally- valid for any orbit (see also chapter 9):

$$F_t = \frac{m \times v^2}{R} \qquad (17.5)$$

In Crenel Physics, it is equation (17.5) that introduces *and* quantifies the appearance 'mass'. In Crenel Physics this equation has macroscopic relevance *and* fundamental (elementary particle) relevance. It is applicable to the Moon orbiting the Earth, and it is applicable for an electron orbiting a nucleus. Here it is postulated that the equation also is valid for the current model. Here, the orbit velocity 'v' is however equal to 'c'. Thus, equation (17.4) can be rewritten as:

$$F_t = \frac{m \times c^2}{R} \qquad (17.6)$$

When the results of equations (17.4) and (17.6) are substituted into equation (17.3), the outcome is:

$$E^+ = R.\,d\alpha \times \frac{m \times c^2}{R} = m.\,c^2.\,d\alpha \qquad (17.7)$$

In this equation parameter 'E^+' represents the energy gain of the child Photon, due to force component 'F_t'.

At the bottom line this amount of energy gain is based on the assumption that energy cannot be lost.

The conclusion is that *any* child photon in the infinite pedigree of parent child successions contains this same amount of energy. And that this energy comes forth from an initial course change '$d\alpha$' that is experienced by an initially free travelling Photon. Once this initial course change has taken place, thereafter an 'energy containing circular orbit of a continued pedigree of parent child successions' is the consequence, whereby per equation (17.7) the associated gain of 'orbit energy' is kept contained equal to $E = m.c^2$. The contained gained energy is proportional to the extent of the initial course change ('dα').

Equation (17.7) quantifies a value for the appearance 'mass' in relation to the appearance 'contained energy'. It actually expresses that the newly introduced variable called 'mass' -introduced into the model by equation (17.5)- is fine, as long as this 'm' obeys the equation $E=m.c^2$. This matches Einstein's famous equation in Metric Physics, and therefore the model at hand confirms this reality.

> *Where the earlier discussed 'Cylinder Model' connected Energy to Frequency (consistent with Planck), the here discussed 'Circular Orbit Model' connects this same Energy to Mass (consistent with Einstein).*

Note that in the 'Circular Orbit Model' the Package containment of the initially free travelling Photon -prior to the course change- seems not relevant to the outcome of the scenario. The gained 'orbit energy' per equation (17.7) is 'as such'.

To further explore the extent of this independency the following analogy can be given:

> Consider some *stone* that is freely travelling through space within some frame of reference. Suddenly, for some reason, this stone is connected to a rope of which the other end is connected to a

fixed point. This will cause the stone to instantaneously start orbiting around this fixed point. As this happens, the stone still remains the same stone, and its forward velocity remains unchanged.

Meanwhile the rope is pulling with a continuous tangential force on the stone, thereby continuously accelerating the stone in a tangential direction. Despite this continuous acceleration, the velocity of the stone remains constant. Furthermore, because the rope length remains constant, the rope does *not* transfer energy to the stone. From a Newtonian viewpoint the energy containment of the system -prior to the stone being connected to the rope- therefore *exactly* matches the situation thereafter.

However, from a Crenel Physics point of view (and also according to Planck's theorem) the now orbiting stone represents *more* energy than the previously free travelling stone.

In Metric Physics:
The difference in energy containment –the energy jump 'E$^+$' caused by the orbit- is according to Planck equal to h.υ. In Metric Physics 'h' is an extremely small number (6.6 x 10^{-34} J.s), and therefore in daily life the difference in energy containment would be unnoticeable. However, should the stone velocity be very high and the rope very short, the orbiting frequency could get really high and the orbit energy 'E$^+$' would become more apparent.

Crenel Physics:

The energy jump 'E$^+$' is expressed by equation (17.7).

In both cases however, the mass of the stone itself is irrelevant to the size of the energy jump.

The analogy between the orbiting Photon and the orbiting stone illustrates, that equation (17.7) expresses in both cases an incremental effect that applies to the orbiting system relative to the non-orbiting system.

18. Comparing Models.

Two models for Photon containment were described until here:

1. The 'Cylinder Model'.
    Here, a Photon is presumed to be trapped in an imaginary cylindrical container with a mirroring internal surface. This model is from a spatial view 1-dimensional.
    The contained energy is according to Planck's equation: $E = h.\upsilon$.

2. The 'Circular Orbit Model'.
    Here, an initially free travelling Photon experiences –for some reason- an initial course change of angle size $d\alpha$. Due to this course change it will appear as if this Photon now starts orbiting (in the model it is in fact an infinitely rapid pedigree of successive parent/child successions). This model is from a spatial view 2-dimensional.
    The orbit itself represents an amount of extra energy ('orbit energy') that is consistent with Einstein's equation: $E = m.c^2$.

This chapter compares both models in Crenel Physics, based on the definition of the Package: an object containing 1 Package has 'a frequency' of 1 cycle per Crenel.

Because in both models it is a Photon that is contained, the velocity of the contained object is 1 Crenel per

Crenel. Therefore, to obtain an object with 1 Package, in the Cylinder Model one needs a container with the length of 0.5 Crenel. Thus, during the timeframe of 1 Crenel the internally bouncing Photon completes one cycle. After each Crenel of time, the original situation has been restored. The amplitude of the movement would be half the container length: 0.25 Crenel.

Likewise, in the 'Circular Orbit Model' the circulating Photon must complete one orbit per Crenel in order to represent the searched for object of 1 Package. The length of the circular orbit therefore must equal 1 Crenel, and therefore the diameter of the orbit must be $1/\pi$ Crenel. Instead of looking at the 'Circular Orbit Model' from a 2-dimensional perspective (as done so far from a North Polestar position), one could position oneself at a remote position *within* the plane of orbit. In doing so, one would –at large distance- 'see' a pendulum (the orbiting Photon) move from left to right, and back again, in a sinus shaped movement where the amplitude equals $1/(2\pi)$ Crenel.

There is an argument to apply the 'Circular Orbit Model' to situations where Package containing entities are shaped (and per Einstein's equation mass and/or energy appears). This model is associated with *rotation*, and thereby with accumulation or spatial containment of rotational energy. While the 'Cylinder Model' would than describe the exclusive gravitational interaction between two of these entities: between two entities one can imagine a Notohp bouncing back and forth between these two entities. As if it were contained within a cylinder.

19. Big Bang.

The previous chapter described how -through some initial course change- a free travelling Photon will end up in an infinitely rapid succession of Photon generations. From the outside world this 'looks as if' the original Photon started orbiting in a confined space. And that it is the orbiting by itself that caused containment of (extra) Packages in the system as a whole, relative to the pre-orbiting situation.

One can now consider an infinite spatial space, in which 'zero-Photons' (as introduced in chapter 5) reside. These 'zero-Photons' have an infinite wavelength, and therefore within this space these 'zero-Photons' would be anywhere at the same time. Because they contain nothing, there is no argument why these 'zero-Photons' could not exist.

> One analogy would be to analyze the liquid surface in a glass filled with water. Once put on a table and come to rest, one could make a statement that 'there are no waves at this surface', the reason being that the surface is flat and remains flat. However, one could also state that there are lots of waves at the surface, but that these are of infinite wavelength. The infinity of this wavelength –combined with a known *finite* wave velocity- causes that the surface is not moving.

> Both statements are equally sound. They both are based on the fact that the surface is not moving.

To undermine the first statement one would have to prove that waves are thinkable at the water surface, and that these waves nevertheless would *not* move the water surface. The aforementioned waves with infinite wavelength are indeed thinkable, which than indeed undermines the first statement.

To undermine the second statement would be more difficult. One therefore would have to prove that the thinkable phenomenon of waves of infinite wavelength could not possibly exist.

Consider the following reasoning:

> At sea one can easily find short waves. The tidal wave caused by the Moon is much longer, and somewhat harder to detect. This tidal wave would however also be present in the aforementioned glass of water. And likewise, the elliptic path of the Earth around the Sun causes an annual wave in that same glass.

> There really is no reason why me myself (being some mass sitting still at the surface of the Earth) is not creating a tidal wave of infinite wavelength at sea… a small one of course.

Thus, the model of an initial space filled with 'zero-Photons' is thinkable. Although these 'zero-Photons' are anywhere at anytime, they still are Photons.

Now assume that just *one* of these Photons would experience a 'course change' as discussed in the previous chapter. According to this model, it would appear as if this single Photon would that start orbiting and thereby equation (17.7) becomes applicable. This particular 'zero-Photon' would then result in a tangible object: a 'newborn object' that contains some non-zero orbiting energy/mass/Packages.

One may expect that at that same moment other 'zero-Photons' –that are anywhere at the same time- would simultaneously collide and interact with the newborn. What then happens may be described as some 'big bang'.